David Starr Jordan

A Catalogue of the Fishes Known to Inhabit the Waters of North America,

north of the Tropic of Cancer, with notes on the species discovered in 1883 and

1884

David Starr Jordan

A Catalogue of the Fishes Known to Inhabit the Waters of North America,
north of the Tropic of Cancer, with notes on the species discovered in 1883 and 1884

ISBN/EAN: 9783337222222

Printed in Europe, USA, Canada, Australia, Japan

Cover: Foto ©Andreas Hilbeck / pixelio.de

More available books at **www.hansebooks.com**

[EXTRACTED FROM THE ANNUAL REPORT OF THE COMMISSIONER OF FISH AND FISHERIES FOR 1884.]

A CATALOGUE

OF THE

FISHES KNOWN TO INHABIT THE WATERS OF NORTH AMERICA,

NORTH OF THE TROPIC OF CANCER,

WITH

NOTES ON THE SPECIES DISCOVERED IN 1883 AND 1884.

BY

DAVID STARR JORDAN.

WASHINGTON:
GOVERNMENT PRINTING OFFICE.
1885.

000.—A CATALOGUE OF THE FISHES KNOWN TO INHABIT THE WATERS OF NORTH AMERICA, NORTH OF THE TROPIC OF CANCER, WITH NOTES ON THE SPECIES DISCOVERED IN 1883 AND 1884.

BY DAVID STARR JORDAN.

The Synopsis of the Fishes of North America, by David S. Jordan and Charles H. Gilbert (Bulletin United S' 'es National Museum No. 16), was finished in September, 1882, and was issued to the public about April 1, 1883.

Since the publication of that work an active study of North American fishes has brought to light many species not included in the Synopsis, and has shown various errors in the nomenclature of species already known. The additions are chiefly in the Bassalian or deep-sea fauna of the Atlantic, in the tropical fauna of the Florida Keys, and in the fresh-water fauna of the lower part of the Mississippi Valley.

It was at first determined to issue these addenda in the form of annual supplements to the Synopsis, but the publication of the supplement for 1883 having been delayed till January, 1885, it has been thought best to unite the lists for 1883 and 1884, and to put the matter in the present form.

I have, therefore, given a list representing the present state of our knowledge of the fishes found north of the Tropic of Cancer, in American waters. In all cases where a species is included which is not in the Synopsis, or in which a name is used in the latter work, different from that here adopted, I have given an explanation, reference or description in the form of a foot-note. Species already fully described elsewhere in publications of the U. S. National Museum are not redescribed here.

In matters of nomenclature and classification I have followed, in this list, the arrangement in the Synopsis, unless important reasons for deviation have appeared. In such cases I have endeavored to avoid premature changes, and the substitution of one doubtful opinion for another.

In this list the families, genera, and species are numbered consecutively from the first. These numbers necessarily differ from those in the Synopsis. The numbers used in that work are here placed in parentheses after the names.

I have also indicated in a general way the geographical distribution of each species by the following signs:

B.—Bassalian or deep-sea fauna of the Atlantic.
BC.—Bassalian fauna of the Pacific.
G.—Arctic (Greenland) fauna.
N.—Shore fauna of North Atlantic States.
S.—Shore fauna of South Atlantic and Gulf States.
W.—West Indian fauna (including Florida Keys).
P.—Tropical fauna of the Pacific coast (Gulf of California to Ecuador).
C.—California shore fauna (Cape Flattery to Cerros Island, &c.).
A.—Alaskan shore fauna.
Y.—Alaskan fresh-water fauna (Yukon).
T.—Fresh-water fauna of region west of Sierra Nevada and Cascade Range (Transmontane).
R.—Fauna of region between Rocky Mountains and Sierra Nevada.
V.—Fresh-water fauna of region east of Rocky Mountains (again subdivided into Vn, the northern part of this range; Vs, the southern; Vsw, the southwestern, &c.)
E.—Europe.
O.—Pelagic species.
Ana.—Anadromous species.
Acc. Accidental visitants.

In this paper I have adopted as the southern boundary of temperate North America the Tropic of Cancer, or a line connecting Key West with Brazos Santiago and Cape San Lucas, instead of the conventional Mexican boundary.

INDIANA UNIVERSITY,
January 1, 1885.

CATALOGUE OF THE FISHES OF NORTH AMERICA.

CLASS I.—LEPTOCARDII. (I)

ORDER A.—CIRROSTOMI. (A)

Family I.—BRANCHIOSTOMIDÆ. (1)

1.—BRANCHIOSTOMA Costa. (1)

1. Branchiostoma lanceolatum Pallas. E. S. C. P. (1)

CLASS II.—MARSIPOBRANCHII. (II)

ORDER B.—HYPEROTRETA. (B)

Family II.—MYXINIDÆ. (2)

2.—MYXINE Linnæus. (2)

2. Myxine glutinosa Linnæus. B. Eu. (2)

Family III.—BDELLOSTOMIDÆ.

3.—POLISTOTREMA Gill. (3)

3. Polistotrema dombeyi Müller. C. (3)

ORDER C.—HYPEROTRETA. (C)

Family IV.—PETROMYZONTIDÆ. (3)

4.—AMMOCŒTES Duméril.[1] (3b.) (4,5)

§ *Entosphenus* Gill. (3b.) (4,5,6)

4. Ammocœtes tridentatus Gairdner. C. Ana. (4)

---

[1] For discussions of the genera of *Petromyzontidæ* see Gill (Proc. U. S. Nat. Mus., 1882, 552) and Jordan & Gilbert (*ibid.*, 1883, 208). Our species fall most naturally into two groups, which we may call genera. *Ammocœtes* with the discal and peripheral teeth differentiated, and the supraoral lamina (maxillary tooth) crescentiform, and *Petromyzon* having the discal and peripheral teeth in obliquely decurved continuous rows, and the supraoral lamina contracted, with 2 or 3 converging teeth. In both groups are minor modifications, indicative of subgenera, the marine species of each (*marinus, tridentatus*) being stronger, with more specialized dentition than the small fluviatile forms.

§ *Lampetra* Gray. (3*pt.*)

5. **Ammocœtes cibarius**[1] Girard. C. Ana. (7)
6. **Ammocœtes aureus** Bean. A. Ana. (7*b*)

§ *Ammocœtes.*

7. **Ammocœtes æpypterus**[2] Abbott. Vn. (8)

5.—PETROMYZON (Artedi) Linnæus. (7)

§ *Ichthyomyzon* Girard. (6)

8. **Petromyzon bdellium**[3] Jordan. Vn. (9)
9. **Petromyzon hirudo** Girard. Vn. (9*b*.)
10. **Petromyzon castaneus** Girard. Vw. (10)

§ *Petromyzon.* (7)

11. **Petromyzon marinus** L. N. Eu. Ana. (11)
11b *Petromyzon marinus dorsatus* Wilder. Ve. (12)

6.—BATHYMYZON[4] Gill.

12. **Bathymyzon bairdii**[5] Gill. B.

Class III.—PISCES.

Subclass ELASMOBRANCHII.

Order D.—OPISTHARTHRI.[6]

Family V.—NOTIDANIDÆ. (15)

7.—HEPTRANCHIAS Ratinesque. (32)

§ *Notorhynchus* Ayres.

13. **Heptranchias maculatus** Ayres. C. (42)

8.—HEXANCHUS Rafinesque. (31*b*.)

14. **Hexanchus corinus** Jordan & Gilbert. C. (42*b*.)

---

[1] The name *Petromyzon plumbeus* is preoccupied by Shaw, 1805.

[2] The name *Petromyzon niger* is preoccupied by Lacépède, 1798. This is probably the species poorly described by Abbott as *Amm. œpyptera*.

[3] The name *Petromyzon argenteus* is preoccupied by Bloch, 1790. I propose the new name *P. bdellium* for this species, as I cannot identify it certainly with *Ammocœtes concolor* Kirtland, *A. borealis* Ag., or any other nominal species, based on larval forms.

[4] BATHYMYZON Gill, Proc. U. S. Nat. Mus., 1883, 254; type *Petromyzon* (*Bathymyzon*) *bairdii* Gill. (βαθυς—deep; μύζω—to suck.) This genus is said to differ from *Petromyzon* in having "the suproral and infroral plates or laminæ destitute of odontoid tubercles, the armature of the lamprey type being obsolescent."

[5] *Petromyzon* (*Bathymyzon*) *bairdii* Gill., l. c. 254, Gulf Stream, latitude 40°, at a depth of 547 fathoms. The species has not been described, except that it is "closely related to *Petromyzon marinus*."

[6] The groups called *Opistharthri* and *Proarthri*, certainly worthy of ordinal distinction from the other Sharks, are defined by Professor Gill in our Synopsis Fish. N. A., 967.

Order E.—PROARTHRI.

Family VI.—CESTRACIIDÆ. (14)

9.—CESTRACION[1] Cuvier. (31)

§ *Gyropleurodus* Gill.

15. Cestracion francisci Girard. C. (41)

Order F.—SQUALI.

Family VII.—SCYMNIDÆ. (4)

10.—ECHINORHINUS Blainville. (8)

16. Echinorhinus spinosus Gmelin. Acc. Eu. (13)

11.—SOMNIOSUS Le Sueur. (9)

17. Somniosus microcephalus Bloch. A. G. Eu. (14)

Family VIII.—SPINACIDÆ. (5)

12.—CENTROSCYLLIUM Müller & Henle. (10)

18. Centroscyllium fabricii Reinhardt. G. (15)

13.—SQUALUS (Artedi) Linnæus. (11)

19. Squalus acanthias Linnæus. C. A. G. N. Eu. (16)

14.—CENTROSCYMNUS Bocage & Capello. (12)

20. Centroscymnus cœlolepis Bocage & Capello. B. Eu. (17)

Family IX.—SCYLLIIDÆ. (6)

15.—SCYLLIORHINUS Blainville. (13*b*.)

§ *Catulus* Smith. (13*b*.)

21. Scylliorhinus ventriosus Garman. C. (18*b*.)
22. Scylliorhinus retifer Garman. B. (18*c*.)

---

[1] CESTRACION Cuvier (Règne Animal, type *Cestracion philippi* Bloch and Schneider) should perhaps be adopted instead of *Heterodontus* Blainville, preoccupied in Herpetology as *Heterodon*. Both words are from ἕτερος, ὀδών (ὀδούς), and are correctly written *Heterodus* or *Heterodon*, not *Heterodontus*. *Cestracion* is an old name of the Hammerheaded shark, from κέστρα, a pick-axe, or similar instrument.

16.—PSEUDOTRIACIS[1] Capello.

23. **Pseudotriacis microdon**[2] Capello. P. En.

17.—GINGLYMOSTOMA Müller & Henle. (13)

24. **Ginglymostoma cirratum** Gmelin. W. P. (18)

Family X.—GALEORHINIDÆ. (7)

18.—GALEUS[3] (Rafinesque) Leach. (14)

§ *Galeus*.

25. **Galeus lunulatus**[4] Jordan & Gilbert. P.

---

[1] PSEUDOTRIACIS Capello. (*Pseudotriakis* Capello, Jorn. Sci. Math. Phys. e Nat. Lisboa, 1868, 321; type *Pseudotriakis microdon* Capello.)

Body elongate; mouth wide, with a very short labial fold near the angle; snout depressed; nostrils inferior, not confluent with the mouth; eyes oblong, lateral, without nictitating membrane; spiracles well developed behind the eye; gill openings moderate, in advance of pectoral; jaws with many rows of very small, tricuspid teeth; first dorsal fin long and low, highest posteriorly, inserted opposite the space between pectorals and ventrals; second dorsal rather large, larger than anal; ventrals and pectorals well developed; no pit at root of caudal; caudal fin divided by a notch into a short upper portion and a very low and long lower portion. Skin with minute asperities. One species known (Ψεῦδος, false; τρειαχις, triacis).

[2] *Pseudotriacis microdon* Capello, Jorn. Sci. Math. &c., Lisboa, 1868, 321; Gunther, VIII, 395; Bean, Proc. U. S. Nat. Mus., VI, 1883, 147. Two specimens of this species are known, the type from Portugal, the second, 10 feet in length, lately taken at Amagansett, on Long Island. (*Bean*.)

[3] GALEUS Rafinesque. (*Mus elus* Cuvier.)

(Rafinesque, Caratteri di alcuni nuovi Generi,1810, 13: *vulpeculus, melastomus, catulus* and *mustelus: Galeus* Leach, Observ. Genus *Squalus* of Linné: 1812, 62, type *Squalus mustelus* Leach = *Sq. canis* Mitchill.)

The name *Galeus* was first used in binomial nomenclature by Rafinesque, for a genus thus defined:

"VIII. G. GALEUS.—Due spiragli, due ale dorsali, un ala anale, cinque branchie da ogni lato; coda diseguale, obliqua.

"*Osservazione.* La maggior parti delli *Squali* degli autori si annoverano in questo genere, il quale si distingue dal vero genere *Squalus* della prezenza di un ala anale."

Four species are mentioned, *vulpeculus: melastomus: catulus* and *mustelus*. Although the species which the author had in mind was probably *Squalus galeus* L., it is improper to assume this species as the type, as no mention is made of it by the author in question.

In 1812, Leach proposed a genus *Galeus*, to include sharks with the anal fin present and the caudal fin irregular (*i. e.*, not lunate). But one species, *Galeus mustelus*, is mentioned by Leach. Still later, a subgenus, *Galeorhinus*, was proposed by Blainville for sharks distinguished from *Carcharinus* Blainv. (=*Carcharias* Cuvier), by the presence of spiracles. In this group are included with others, *Squalus mustelus* and *Squalus galeus* of Linnæus. Still later (1817), the genera *Mustelus, Carcharias,* and *Galeus* were defined by Cuvier, and with his definition have been accepted by nearly all later authors.

The rules of nomenclature seem to me to require the retention of the genus *Galeus* Rafinesque, for the group for which the same name was used by Leach, *i e.*, instead of *Mustelus* Cuvier.

[4] *Mustelus lunulatus* Jordan & Gilbert, Proc. U. S. Nat. Mus., 1882, 108; Mazatlan, Mexico.

In this paper is given an analysis of the distinctive characters of the four North American species of *Galeus*:—*lunulatus, canis, dorsalis,* and *californicus*.

[7] CATALOGUE OF THE FISHES OF NORTH AMERICA.

26. Galeus canis Mitchill. N. Eu. (19)

§ *Pleuracromylon* Gill.

27. Galeus californicus Gill. C.

19.—TRIACIS Müller & Henle. (15)

§ *Triacis.*

28. Triacis semifasciatus Girard. C. (21)

§ *Rhinotriacis* Gill.

29. Triacis henlei Gill. C. (22)

20.—GALEORHINUS Blainville. (16)

30. Galeorhinus zyopterus Jordan & Gilbert. C. (23)

21.—GALEOCERDO Müller & Henle. (17)

31. Galeocerdo maculatus[1] Ranzani. W. P. (24)

22.—CARCHARHINUS[2] Blainville. (18, 19, 20, 21)

§ *Carcharinus.*

32. Carcharhinus glaucus Linnæus. C. O. Eu. (25)

§ *Eulamia* Gill.

33. Carcharhinus obscurus Le Sueur. N. (26)
34. Carcharhinus æthalorus[3] Jordan & Gilbert. P.
35. Carcharhinus fronto[4] Jordan & Gilbert. P.
36. Carcharhinus platyodon Poey. W. S. (26b.)

---

[1] *Galeus maculatus* Ranzani, De Novis Speciebus Piscium, Dissert. Prima, 1838, 7; *Galeocerdo maculatus*, Poey, Enumeratio Pisc. Cubens., 201, 1875. This name has priority over *G. tigrinus* Müller & Henle.

[2] Although *Carcharias glaucus* was probably the species in mind when Rafinesque proposed his genus *Carcharias*, he makes no reference to this species. The only species actually mentioned by him in connection with the original account of his genus *Carcharias* is *Odontaspis taurus*. The name *Carcharias*, if used at all, should supersede *Odontaspis*. This is the view at first taken by us in the Synopsis Fish. N. A., but afterwards, in the Addendum, p. 872, changed to follow current usage.

The oldest tenable name of this group is that of *Carcharhinus* Blainville. I think it best to regard *Eulamia*, *Aprionodon*, *Hypoprion*, and *Scoliodon* as subgenera under *Carcharhinus*, rather than as distinct genera.

[3] *Carcharias æthalorus* Jordan & Gilbert, Proc. U. S. Nat. Mus., 1882, 104; Mazatlan: Panama.

[4] *Carcharias fronto* Jordan & Gilbert, Proc. U. S. Nat. Mus., 1882, 102. Mazatlan.

37. **Carcharhinus caudatus**[1] De Kay. N. (27)
38. **Carcharhinus lamia**[2] Risso. W. Eu.
39. **Carcharhinus lamiella** Jordan & Gilbert. C. (27b.)

§ *Hypoprion* Müller & Henle. (19b)

40. **Carcharhinus brevirostris**[3] Poey. W. (28b.)

§ *Isogomphodon* Gill. (19)

41. **Carcharhinus limbatus** Müller & Henle. W. Acc. (28)

§ *Aprionodon* Gill.

42. **Carcharhinus isodon**[4] Müller & Henle. W. Acc. (29)

§ *Scoliodon* Müller & Henle. (21)

43. **Carcharhinus longurio**[5] Jordan & Gilbert. P.
44. **Carcharhinus terræ-novæ**[6] Richardson. N. S. W. (30)

Family XI.—SPHYRNIDÆ. (8)

23.—**SPHYRNA** Rafinesque. (22, 23)

§ *Reniceps* Gill. (22)

45. **Sphyrna tiburo** Gill. S. W. (31)

---

[1] The name *cæruleus* is preoccupied in this genus by the *Squalus* (*Carcharinus*) *cæruleus* of Blainville, 1816, a synonym of *Carcharhinus glaucus*. The name next in date is that of *Lamna caudata* De Kay, New York Fauna, Fishes, 1842, 354.

[2] *Carcharhinus lamia*. This species is described on page 873, in the Synopsis. It is abundant in the Mediterranean and in the West Indies, ranging northward to the Florida Keys, being common about the wharves at Key West. Base of first dorsal 1⅞ in interspace between dorsals; base of second, 4¼; length of pectoral, about 5 in length of body.

(*Carcharias lamia* Rafinesque, Indice, 1810, 44; name only; *Squalus carcharias* (in part?) Cuvier (Règne Animal), and of several authors; not of Linnæus; *Carcharias lamia* Risso, Hist. Nat. Europ. Mérid., III, 119, 1826; *Squalus longimanus* Poey, Memorias Cuba, II, 338; *Eulamia longimana* Poey, Syn. Pisc. Cubens., 1868, 448; *Eulamia lamia* Poey, Enum. Pisc. Cubens., 188; *Carcharias lamia* Jordan, Proc. U. S. Nat. Mus., 1884, 104 (Key West).)

[3] *Carcharhinus brevirostris* is described in detail by Jordan & Gilbert, Proc. U. S. Nat. Mus., 1882, 581, and by Jordan *op. cit.*, 1884, 104, from specimens obtained at Charleston and Key West.

[4] *Carcharhinus isodon*, briefly described in the Synopsis (p. 24) as *Aprionodon punctatus*, is a West Indian species, very lately obtained for the first time on our coast. (Parker.)

[5] *Carcharias longurio* Jordan & Gilbert, Proc. U. S. Nat. Mus., 1882, 106; Mazatlan.

[6] Specimens of *Scoliodon terræ-novæ*, *Malthe radiata* (*cubifrons*), *Scorpæna plumieri* (*bufo*), and other fishes of the warm seas, were given by Audubon to Richardson, and by Richardson described as coming from the waters about Newfoundland. There can be little doubt that these specimens really came from Southern Florida, in which region Audubon made extensive collections. The *Squalus punctatus* of Mitchill has been identified by me with *C. terræ-novæ*, and by Prof. Gill with *C. isodon*. The name *punctatus* is any case preoccupied and cannot be used for either species. *Squalus punctatus* Bloch & Schneider-1801, is a *Ginglymostoma*.

§ *Sphyrna.*

46. **Sphyrna tudes**[1] Cuvier. W. P. Eu.
47. **Sphyrna zygæna** Linnæus. N. S. W. C. P. (32)

Family XII.—ALOPIIDÆ. (9)

24.—ALOPIAS Rafinesque. (24)

48. **Alopias vulpes** Gmelin. C. N. Eu. (33)

Family XIII.—ODONTASPIDIDÆ. (10).

25.—CARCHARIAS[2] Rafinesque. (25)

§ *Eugomphodus* Gill.

49. **Carcharias littoralis** Mitchill. N. (34)

Family XIV.—LAMNIDÆ. (11)

26.—ISURUS Rafinesque. (26)

§ *Isuropsis* Gill.

50. **Isurus dekayi** Gill. W. S. (35; 36)

27.—LAMNA Cuvier. (27)

51. **Lamna cornubica** Gmelin. C. Eu. N. (37)

28.—CARCHARODON Smith. (28)

52. **Carcharodon carcharias**[3] Linnæus. C. N. Eu. O. (38)

Family XV.—CETORHINIDÆ. (12)

29.—CETORHINUS Blainville. (29)

53. **Cetorhinus maximus** Gunner. C. N. Eu. O. (39)

---

[1] *Sphyrna tudes* Cuvier. Intermediate in all respects between *S. zygæna* and *S. tiburo*, the hammer longer and less produced laterally than in the former. Anterior margin of the head much curved, but not continuous with the lateral edge; length of hinder margin of one side of the hammer less than its width near the eye. Nostril close to the eye, its groove longer than in *S. tiburo*, but very short, continued for but a short distance along the side of the head, and followed by a line of pores.

A large shark, of the warm seas, Gulf of California, West Indies, Mediterranean, and Indian Ocean.

(*Zygæna tudes* Cuvier (Règne Animal); *Sphyrna tudes* Müller & Henle, Plagiost., 53; *Zygæna tudes* Günther, VIII, 382; *Sphyrna tudes* Jordan & Gilbert, Bull. U. S. Fish Comm., 1882, 105.)

[2] *Carcharias* Rafinesque was established for those sharks, "the most enormous and most voracious of their order, which differ from the genus *Galeus* Rafinesque, by the lack of spiracles." But one species (*Carcharias taurus* Rafinesque) is mentioned, and this species, although really possessing spiracles, must be regarded as the type of *Carcharias*. This name should therefore supersede *Odontaspis*.

[3] A good account of this species is given by Dr. W. B. Stevenson, Proc. Vassar Brothers Sci. Soc., Poughkeepsie, 1884, and in American Naturalist for the same year.

Family XVI.—RHINODONTIDÆ. (13)

30.—MICRISTODUS Gill. (30)

54. **Micristodus punctatus** Gill. P. (40)

Family XVII.—SQUATINIDÆ. (16)

31.—SQUATINA Duméril. (33)

55. **Squatina squatina**[1] Linnæus. C. N. Eu. (43)

Order G.—RAIÆ. (E)

Family XVIII.—PRISTIDIDÆ. (17)

32.—PRISTIS. Latham. (34)

56. **Pristis pectinatus** Latham. W. S. (44)
57. **Pristis perrottetii**[2] Müller & Henle. P.

Family XIX.—RHINOBATIDÆ. (18)

33.—RHINOBATUS Bloch & Schneider. (35)

§ *Rhinobatus.*

58. **Rhinobatus productus** Ayres. C. (45)
59. **Rhinobatus glaucostigma**[3] Jordan & Gilbert. P.
60. **Rhinobatus lengtiginosus** Garman. W. (45d)

§ *Zapteryx.* Jordan & Gilbert.

61. **Rhinobatus exasperatus** Jordan & Gilbert. C. P. (45b)

§ *Platyrhinoidis.* Garman.

62. **Rhinobatus triseriatus** Jordan & Gilbert. C. (45c)

---

[1] Our reasons for retaining the original specific name, even when identical with the name of the genus, have been given in full in Proc. U. S. Nat. Mus., 1884, 18. The same view of the case has been adopted by the American Ornithologists' Union.

[2] *Pristis perrotteti* Müller & Henle. Rostral teeth in 18 or 20 pairs, not trenchant behind; distant from one another, the base of each tooth being about one-third the interspaces. Dorsal fin nearly in advance of ventrals. Root of pectoral in advance of first gill-opening, its outer angle a right one. Second dorsal not much smaller than first; a smaller lower caudal lobe. (Günther.) Tropical seas, north to Mazatlan, on the Pacific coast.

(Müller & Henle, 108; Günther, VIII, 436; Jordan & Gilbert, Bull. U. S. Nat. Mus., 1882, 105.)

[3] *Rhinobatus glaucostigma* Jordan & Gilbert, Proc. U. S. Nat. Mus., 1883, 210. Mazatlan; Gulf of California.

Family XX.—RAIIDÆ. (20)

34.—RAIA Linnæus. (37)

63. Raia erinacea Mitchill. N. (48)
64. Raia ocellata Mitchill. N. (49)
65. Raia radiata Donovan. N. Eu. (50)
66. Raia eglanteria Lacépède. N. (51)
67. Raia ackleyi ornata Garman. W. B. (53o.)
68. Raia plutonia Garman. W. B. (53o.)
69. Raia granulata Gill. B. (53)
70. Raia parmifera Bean. A. (57b.)
71. Raia stellulata Jordan & Gilbert. C. (57)
72. Raia inornata Jordan & Gilbert. C. (56)
72 b *Raia inornata inermis* Jordan & Gilbert. C.
73. Raia rhina Jordan & Gilbert. C. A. (55)
74. Raia binoculata Cooper. C. A. (54)
75. Raia lævis Mitchill. N. (52)

Family XXI.—TORPEDINIDÆ. (19)

35.—TORPEDO Duméril. (36)

76. Torpedo occidentalis Storer. E. (46)
77. Torpedo californica Ayres. W. (47)

36.—NARCINE Müller & Henle. (36b.)

78. Narcine brasiliensis Olfers. W. (47b.)
78 b *Narcine brasiliensis corallina* Garman. W.
79. Narcine umbrosa[1] Jordan. W.

Family XXII.—TRYGONIDÆ. (21)

37.—UROLOPHUS Müller & Henle. (38)

80. Urolophus halleri Cooper. C. P. (58)
81. Urolophus asterias[2] Jordan & Gilbert. P.

38.—PTEROPLATEA Müller & Henle. (39)

82. Pteroplatea crebripunctata[3] Peters. P.
83. Pteroplatea maclura Le Sueur. S. (59)
84. Pteroplatea marmorata Cooper. C. (60)

---

[1] *Narcine umbrosa* Jordan, Proc. U. S. Nat. Mus., 1884, 105 ; Key West.

[2] *Urolophus asterias* Jordan & Gilbert, Proc. U. S. Nat. Mus., 1882, 579 ; Mazatlan, Panama.

[3] *Pteroplatea crebripunctata* Peters, Monatsber, Beri. Akad, 1869, 703. This species is very common in the Gulf of California. It is thus described by Dr. Peters:

Breadth of disk twice the distance from tip of snout to vent. Snout with a blunt projection ; anterior margin of pectorals undulate, convex anteriorly and posteriorly, medially weakly concave ; outer angle sharply rounded ; posterior margins weakly convex, the posterior angle rounded, covering outer half of base of ventrals ; spiracle without tentacle ; tail (mutilated) with a low fold on its upper edge. Brown above, with thick-set black points ; a row of small, close-set yellow spots on front of disk ; under side yellowish.

I have compared specimens of this species with *P. maclura* and *P. marmorata*, and regard the three as unquestionably distinct, although closely related.

39.—TRYGON Adanson. (40)

85. **Trygon centura** Mitchill. N. (61)
86. **Trygon hastata** De Kay. S. (62b)
87. **Trygon sayi** Le Sueur. S. W. (62)
88. **Trygon longa**[1] Garman. P.
89. **Trygon dipterura** Jordan & Gilbert. C. (63)
90. **Trygon tuberculata** Lacépède. W. (64)
91. **Trygon sabina** Le Sueur. S. (65)

Family XXIII.—MYLIOBATIDÆ. (22.)

40.—STOASODON Cantor. (41)

92. **Stoasodon narinari** Euphrasen. S. W. (66)
93. **Stoasodon laticeps**[2] Gill. P.

41.—MYLIOBATIS Duméril. (42)

94. **Myliobatis freminvillei** Le Sueur. E. S. (67)
95. **Myliobatis californicus** Gill. C. (68)

42.—RHINOPTERA Kuhl. (43)

96. **Rhinoptera quadriloba** Le Sueur. N. (69)

Family XXIV.—CEPHALOPTERIDÆ. (23.)

43.—MANTA Bancroft. (44)

97. **Manta birostris** Walbaum. S. P. W. (70)

Subclass HOLOCEPHALI.

Order H.—HOLOCEPHALI. (F)

Family XXV.—CHIMÆRIDÆ. (24)

44.—CHIMÆRA Linnæus. (45)

§ *Chimæra.*

98. **Chimæra affinis** Capello.[3] B. Eu. (71)

§ *Hydrolagus* Gill.

99. **Chimæra colliei** Bennett. C. A. (72)

---

[1] *Trygon longa* Garman. This species is described in the Synopsis Fish N. A., p. 66. It is not uncommon along the Pacific coast, from the Gulf of California to Panama.

[2] *Aëtobatis laticeps* Gill, Ann. Lyc. Nat. Hist. N. Y., 1865, 137. This species is abundant from the Gulf of California southward. It has never been properly compared with *S. narinari*, and may not be different.

[3] *Chimæra plumbea* and *abbreviata* Gill.

To the synonymy in the Synopsis (. 54) add: *Chimæra affinis* Capello, Jorn. Sci. Math. Phys. e. Nat., Lisboa, IV, 1868, 314, pl. III (facing p. 274), ff. 1, 1a.; Günther, VIII, 350; *Chimæra abbreviata* Gill, Proc. U. S. Nat. Mus., 1883, VI, 254.)

We are indebted to Dr. Bean for the information that the *Chimæra plumbea* and *Chimæra abbreviata* of Dr. Gill are identical with each other and with *Ch. affinis*.

Subclass ACTINOPTERI.

Order I.—SELACHOSTOMI. (G)

Family XXVI.—POLYODONTIDÆ. (25)

45.—POLYODON Lacépède. (46)

100. **Polyodon spathula** Walbaum. Vw. (73)

Order J.—GLANIOSTOMI. (H)

Family XXVII.—ACIPENSERIDÆ. (26)

46.—ACIPENSER Linnæus. (47)

101. **Acipenser sturio oxyrhynchus** Mitchill. N. Ana. (74).
102. **Acipenser transmontanus** Richardson. C. A. Ana. (75)
103. **Acipenser medirostris** Ayres. C. A. Ana. (76)
104. **Acipenser rubicundus** Le Sueur. Vn. (77)
105. **Acipenser brevirostris** Le Sueur. N. S. (78)

47.—SCAPHIRHYNCHOPS Gill. (48)

106. **Scaphirhynchops platyrhynchus** Rafinesque. Vw. (79)

Order K.—GINGLYMODI.[1] (I)

Family XXVIII.—LEPIDOSTEIDÆ. (27)

48.—LEPIDOSTEUS Lacépède. (50)

107. **Lepidosteus osseus** Linnæus. V. (80)
108. **Lepidosteus platystomus** Rafinesque. V. (81)
109. **Lepidosteus tristœchus**[2] Bloch & Schneider. Vs. W. (82)

Order L.—HALECOMORPHI. (J)

Family XXIX.—AMIIDÆ. (28)

49.—AMIA Linnæus. (51)

110. **Amia calva** Linnæus. V. (83)

---

[1] The word *Ginglymodi* is from γιγγλυμὸς, hinge, ειδος, like, in allusion to the ball-and-socket joints of the vertebræ.

[2] The subdivisions of *Lepidosteus* (*Cylindrosteus; Atractosteus*) certainly have no value higher than specific, and the characters used in distinguishing them are variable and of slight importance. It is often difficult to distinguish *L. platystomus*, even specifically, from *L. tristœchus*. Specimens from Cuba (*tristœchus*) are not distinguishable from others from Florida (*spatula*).

Order M.—NEMATOGNATHI. (K)

Family XXX.—SILURIDÆ. (29)

50.—NOTURUS Rafinesque. (52)

§ *Schilbeodes* Bleeker.

111. Noturus gyrinus Mitchill. Vn. (84)
112. Noturus leptacanthus Jordan. Vs. (85)
113. Noturus nocturnus[1] Jordan & Gilbert. Vw.
114. Noturus funebris[2] Gilbert & Swain. Vs.
115. Noturus latifrons[3] Gilbert & Swain. Vc.
116. Noturus miurus[4] Jordan. V. (86, 87)
117. Noturus exilis[5] Nelson. Vw. (88)
118. Noturus insignis Richardson. Vc. (89)

§ *Noturus*.

119. Noturus flavus Rafinesque. Vw. (90)

51.—LEPTOPS Rafinesque. (53)

120. Leptops olivaris Rafinesque. V. (91)

52.—GRONIAS Cope. (54)

121. Gronias nigrilabris Cope. Ve. (92)

53.—AMIURUS Rafinesque. (55)

122. Amiurus brunneus Jordan. Vsc. (93)
123. Amiurus platycephalus Girard. Vse. (94)
124. Amiurus melas[6] Rafinesque. Vw. (95, 96)
125. Amiurus nebulosus[7] Le Sueur. V. (98)
125 b. *Amiurus nebulosus catulus*[8] Girard. Vsw.

---

[1] *Noturus nocturnus* Jordan & Gilbert, Proc. U. S. Nat. Mus., 1885. Arkansas to Texas.

[2] *Noturus funebris* Gilbert & Swain, Proc. U. S. Nat. Mus., 1885. Northern Alabama.

[3] *Noturus latifrons* Gilbert & Swain, Proc. U. S. Nat. Mus., 1885. White River, Indiana.

[4] *Noturus eleutherus* seems to be inseparable from *Noturus miurus*.

[5] *Noturus elassochir* Swain & Kalb (Proc. U. S. Nat. Mus., 1882, 639) seems to me identical with *Noturus exilis*. I regard the latter as distinct from *N. insignis*. For a detailed review of the genus *Noturus*, see Swain & Kalb, *loc cit*.

[6] The species called in the Synopsis *Amiurus xanthocephalus* seems to be not distinct from *A. melas*. *Amiurus cragini* Gilbert, Bull. Washburn Lab. Nat. Hist., 1884, 1, 10, from Kansas, is identical with *Amiurus obesus* Gill, which I regard as the original *melas* of Rafinesque. *Amiurus brachyacanthus* Cope is probably the same species. The chief characters by which *A. melas* is distinguished from *A. nebulosus* are the much shorter pectoral spines and shorter anal fin of the former.

[7] The original *Silurus catus* L. was certainly not this species, or any other North American siluroid. The oldest tenable specific name for this species is that of *nebulosus* Le Sueur.

[8] The type of *Pimelodus catulus* Girard should be referred to *A. nebulosus* rather than to *A. melas*. It represents a slight variety of *A. melas* occurring in the lower Mississippi Valley and Texas.

[15] CATALOGUE OF THE FISHES OF NORTH AMERICA.

125c. *Amiurus nebulosus marmoratus*[1] Holbrook. Vs. (97)
126. **Amiurus vulgaris** Thompson. Vn. (99)
127. **Amiurus natalis** Le Sueur. V. (100)
127b. *Amiurus natalis lividus* Rafinesque. V.
127c. *Amiurus natalis bolli* Cope. Vsw. (100b.)
128. **Amiurus erebennus**[2] Jordan. Vse. (101)
129. **Amiurus albidus**[3] Le Sueur. Ve. (102, 103)
130. **Amiurus lupus** Girard. Vsw. (104)
131. **Amiurus niveiventris** Cope. Vse. (105)
132. **Amiurus nigricans** Le Sueur. Vw. (106)
133. **Amiurus ponderosus**[4] Bean. Vw. (107)

54.—**ICTALURUS**[5] Rafinesque. (56)
134. Ictalurus punctatus Rafinesque. V. (108)
135. Ictalurus furcatus Cuv. & Val. Vsw. (109)

55.—**GALEICHTHYS**[6] Cuv. & Val. (57)
§ *Arius* Cuv. & Val.
136. Galeichthys guatemalensis[7] Günther. P.
137. Galeichthys seemanni[8] Günther. P.

---

[1] *Amiurus marmoratus* represents apparently a color variety only of *Amiurus nebulosus*. It inhabits grassy waters southward.

[2] Professor Cope describes (Proc. Ac. Nat. Sci., Phila., 1883, 133) a catfish from Batstoe River, New Jersey, as a new species, under the name of *Amiurus prosthistius*. Except that the caudal fin is said to be rounded rather than truncate, this species agrees with *A. erebennus*, with which species we think that it will prove identical. Greatest width of head equal to depth of body; eye small, 5 in interorbital width; dorsal spine inserted much nearer tip of snout than adipose fin; pectoral spines a little larger than dorsal spine; maxillary barbel reaching middle of pectoral spine; humeral process extending a little farther; black, whitish below; fins black; pectoral and ventral pale at base; head, $3\frac{2}{3}$; depth, $4\frac{1}{4}$. D. I. 6. A. 24 to 27. Batstoe River, New Jersey. (*Cope.*)

[3] *Amiurus lophius* Cope seems to be the adult form of *A. albidus*.

[4] *Amiurus ponderosus* is perhaps the adult form of *A. nigricans*. The type of the former species has 35 anal rays. We have counted 25, 27, 28, and 32 anal rays in four individuals of *A. nigricans*.

[5] It is probably better, if the genus *Amiurus* is to be retained as distinct from *Ictalurus*, to refer to it all the transitional species having the tail forked and the bony bridge, from occiput to dorsal not quite continuous. It is true that this latter character is largely one of degree, but still there is a positive difference between *I. punctatus* and *furcatus* and the fork-tailed *Amiuri*.

[6] GALEICHTHYS Cuvier & Valenciennes.
Arius (C. & V.); *Hexanematichthys, Guiritinga, Hemiarius, Cephalocassis, Netuma,* and *Pseudarius* Bleeker; *Notarius, Ariopsis,* and *Leptarius* Gill; *Sciadarius* and *Bagropsis* Kner; *Cathorops* Jor. & Gilb.).
(Cuvier & Valenciennes, Hist. Nat. Poiss., XV., 29, 1840; type *Galeichthys feliceps* C. & V.).
The genus *Arius*, distinguished from *Galeichthys* by having the nuchal shield ("occipital process") not covered by thick skin, cannot well be separated from *Arius*, as in several species (*dasycephalus, brandti* &c.) this character is simply sexual. For a full account of the species of this genus, found on the west coast of America, see Jordan & Gilbert, Bull. U. S. Fish Comm., 1882, 34.

[7] *Arius guatemalensis* Günther, V. 1864, 145; Jordan & Gilbert, Bull. U. S. Fish Comm., 1882, 48; Mazatlan to Panama.

[8] *Arius seemanni* Günther, V. 147; *Arius assimilis* Jordan & Gilbert, Bull. U. S. Fish Comm., 1882, 47 (not *A. assimilis* Günther); Mazatlan to Panama.

138. Galeichthys felis Linnæus. N. S. (110, 111)
139. Galeichthys platypogon[1] Günther. P.
140. Galeichthys brandti[2] Steindachner. P.

56.—ÆLURICHTHYS Baird & Girard. (58)

141. Ælurichthys marinus Mitchill. S. (112)
142. Ælurichthys panamensis[3] Gill. P.
143. Ælurichthys pinnimaculatus[4] Steindachner. P.

Order N.—EVENTOGNATHI. (L)

Family XXXI.—CATOSTOMIDÆ. (30)

57.—ICTIOBUS Rafinesque. (59, 60, 61)

§ *Sclerognathus* Cuv. & Val. (59)

144. Ictiobus cyprinella Cuv. & Val. Vw. (113)

§ *Ictiobus*. (60)

145. Ictiobus urus Agassiz. Vw. (114)
146. Ictiobus bubalus Rafinesque. Vw. (115)

§ *Carpiodes* Rafinesque. (61)

147. Ictiobus carpio[5] Rafinesque. Vw. (116)
148. Ictiobus velifer[6] Rafinesque. Vw. (120)
148 b. *Ictiobus velifer bison* Agassiz. Vw. (119)
148 c. *Ictiobus velifer tumidus* Baird & Girard. wV. (117)

---

[1] *Arius platypogon* Günther, V. 147; Jordan & Gilbert, Bull. U. S. Fish Comm., 1882, 44; Mazatlan to Panama.

[2] *Arius brandti* Steindachner, Ichthyol, Beitr., IV, 21, 1875; Jordan & Gilbert, Bull. U. S. Fish Comm., 1882, 39; Mazatlan to Panama.

[3] *Ælurichthys panamensis* Gill. Proc. Ac. Nat. Sci., Phila., 1863, 172 = *Ælurichthys nuchalis* Günther, V, 179, 1865 = *Ælurichthys panamensis* Jordan & Gilbert. Bull. U. S. Fish Comm., 1882, 35; Mazatlan to Panama.

[4] *Ælurichthys pinnimaculatus* Steindachner, Ichth., Beitr., IV, 15, 1875, Jordan & Gilbert, Bull. U. S. Fish Comm., 1882, 34; Mazatlan to Panama.

[5] This species is very distinct from the others referred to *Carpiodes*. Its body is almost fusiform, the depth about 3 times in length, the head 4⅔, and the first ray of the dorsal not more than half the length of the base of the fin.

[6] Excepting *I. carpio*, all the other specimens of *Carpiodes* which I have examined from points west of the Allegheny Mountains seem to me to belong to a single extremely variable or polymorphous species, *I. velifer*. As varieties, we may perhaps recognize *tumidus* (= *grayi*), with high back and small eye; *bison* (= *damalis*), with large eye, moderate fins, and snout little obtuse; *velifer*, with snout little obtuse, and the dorsal fin very high, and *difformis*, with very blunt snout, large eye, and very high fins. These forms, however, appear to intergrade perfectly.

148 d. *Ictiobus velifer difformis* Cope. Vw. (121)
149. **Ictiobus cyprinus**[1] Le Sueur. Ve.

58.—CYCLEPTUS Rafinesque. (62)

150. **Cycleptus elongatus** Le Sueur. Vw. (122)

59.—PANTOSTEUS Cope. (63)

151. **Pantosteus plebeius**[2] Baird & Girard. R. (123, 124, 125)
152. **Pantosteus generosus**[3] Girard. R. (126, 127)
153. **Pantosteus guzmaniensis**[4] Girard. R. (128)

60.—CATOSTOMUS Le Sueur. (64)

154. **Catostomus aræopus** Jordan. T. (134)
155. **Catostomus clarki**[5] Baird & Girard. R. (144)
156. **Catostomus discobolus** Cope. R. (129)
157. **Catostomus latipinnis** Baird & Girard. R. (130)
158. **Catostomus nebulifer** Garman. R. (130c.)
159. **Catostomus retropinnis** Jordan. R. (130)
160. **Catostomus catostomus**[6] Forster. Vn. Y. (132)
161. **Catostomus tahoensis** Gill & Jordan. R. (133)
162. **Catostomus labiatus** Ayres. T. (133)
163. **Catostomus macrochilus** Girard. T. (136)
164. **Catostomus occidentalis** Ayres. T. (137)

---

[1] All the specimens of *Carpiodes* from east of the Allegheny Mountains examined by me belong to a species closely related to *I. velifer*, but with the opercle nearly smooth, instead of strongly striate, as in the western species. In the eastern form, *I. cyprinus*, the eye is quite small, the body rather deep, and the dorsal fin rather high.

[2] *Pantosteus bardus* and *delphinus* are almost certainly identical with *P. plebeius*. The type of the latter species has the scales 90-30, less crowded forwards than in *P. generosus*; those before the dorsal much less reduced in size. Dorsal rays, 9; head, 4⅔; depth, 5: snout moderately broad, projecting; fins much lower than in *P. guzmaniensis*.

[3] *Pantosteus platyrhynchus* is based on shriveled specimens of *P. generosus*.

[4] The type of *Catostomus guzmaniensis*, lately examined by me, is a *Pantosteus*, and I am unable to distinguish it from the type of *P. virescens* on comparison of the two specimens. Lat. l. 100 in *guzmaniensis*. Scales before dorsal, 46 to 53; fins high.

[5] The type of *Catostomus clarki*, lately found, belongs to a species very closely related to *C. aræopus*, having the restricted fontanelle and cartilaginous lips of the latter species, but with the scales less crowded anteriorly, there being but 23 in a line before the dorsal instead of 42, as in *C. aræopus*. D. 11; lat. l. 70. *C. discobolus*, *C. aræopus*, and *C. clarki* mark a transition from *Catostomus* toward *Pantosteus*.

[6] Called in the text, *Catostomus longirostris*. The form described by Mr. Mather under the name of *Catostomus nanomyzon* should apparently be referred to this species. Brown; male with a red lateral band in the breeding season; head slender, flattened above; the snout shorter than in *C. catostomus*; lips thick, the lower with 3 or 4 rows of tubercles; eye large, 4 in head, 1½ in snout. Scales smaller anteriorly, but little crowded; dorsal higher than long; pectorals reaching front of dorsal; head, 4; depth, 5; D. 1, 10; A. 7; V. 9; scales, 14-99-11; L. (spawning specimens) 4¼ inches. Big Moose Lake, Adirondack region. Apparently a dwarfed brook variety of *C. catostomus*, but inhabiting the same region and spawning at a much smaller size. (*Mather*.) (*Catostomus nanomyzon*, Twelfth Rept. Survey Adirondack Region, 1884, 36.)

165. Catostomus bernardini[1] Girard. T. (138)
166. Catostomus ardens Jordan & Gilbert. R. (139)
167. Catostomus fecundus Cope & Yarrow. R. (140)
168. Catostomus cypho Lockington. R. (141)
169. Catostomus insignis[2] Baird & Girard. E. (142)
170. Catostomus teres[3] Mitchill. R. (143)

61.—HYPENTELIUM[4] Rafinesque.

171. Hypentelium nigricans Le Sueur. Vw. (145)

62.—CHASMISTES Jordan. (65)

172. Chasmistes liorus Jordan. R. (146)
173. Chasmistes brevirostris Cope T. (147)
174. Chasmistes luxatus Cope. T. (148)
175. Chasmistes cujus[5] Cope. R.

[1] The type of *Catostomus bernardini* is closely related to *C. occidentalis*, differing chiefly in the less conic form of the head and in the larger lower fins. Scales much crowded forwards; 31 before the dorsal (40 in *C. occidentalis*), 75 in the lateral line. Fontanelle large; lips broad, without cartilaginous sheath, formed as in *C. occidentalis*, the lower deeply incised; fins high, the dorsal longer than high, with 12 rays; caudal lobes equal; head 4¼ in length.

[2] *Catostomus insignis* (type lately found) is closely related to *C. teres*, differing chiefly in the broader upper lip, which has several rows of tubercles upon it. Fontanelle rather small; no cartilaginous sheath on lower lips; scales considerably crowded anteriorly, much more so than in *C. clarki*; 27 scales before dorsal; 56 in lateral line. D. 11.

[3] Called in the text, *Catostomus commersoni*. Although the *Cyprinus commersoni* of Lacépède is probably a sucker and may be this species, there is no certainty in so identifying it, the description being very imperfect and the type said to have been observed by Commerson in the East Indies; a statement apparently derived from a confusion of manuscripts of Commerson with those of Bosc. We think it better to retain for this species the later name of *teres*, concerning which no doubt exists. To this species apparently should be referred the small "June sucker" of the Adirondacks, described by Mather as *Catostomus utawana*. Olivaceous, white below; males without red in the breeding season; body slender; head not small, flattened above; snout little prominent; upper lip with two rows of papillæ; eye 4 in head; 2 in snout; dorsal as long as high; pectorals nearly reaching front of dorsal; head 4; D. 1, 11; A. 5; V. 9. Scales 9–67–8; length of adult 4½ inches. Blue Mountain Lakes, Adirondack region. (*Mather*.) Apparently a mountain race of *C. teres*. (Mather. Twelfth Rept., Survey Adirondack Region, N. Y., 35.)

"This small fish I was at first disposed to consider as a dwarfed mountain form of *C. commersoni*, but the fact that the latter fish is found in waters inhabited by this species, and while it grows to a length of 12 or more inches there, this little sucker barely reaches five. Added to this the fact that the larger species had finished spawning in the inlets in May, while this fish was found in masses in the swift mountain streams which tumble rapidly over rocks in the latter part of June, depositing their eggs, thereby showing that they are adult fish." (*Mather*.)

[4] In view of the peculiar form of the cranium in *Catostomus nigricans*, contrasting with that seen in all the other *Catostominæ*, it is probably well to regard it as the type of a distinct genus, *Hypentelium* Rafinesque.

[5] *Chasmistes cujus* Cope. *Couia*.

Pale olive; head broad and flat; upper lip very thin; lower lip represented by folds on each side, which do not connect around the symphysis; eye 8½ in head; in-

63.—ERIMYZON Jordan. (66)

176. Erimyzon sucetta[1] Lacépède. Vs. (150)
176 b. *Erimyzon sucetta oblongus* Mitchill. Vn. (149)

64.—MINYTREMA Jordan. (67)

177. Minytrema melanops Rafinesque. Vw. (151)

65.—MOXOSTOMA Rafinesque. (68)

178. Moxostoma papillosum Cope. Vse. (152)
179. Moxostoma velatum Cope. Vw. (153)
180. Moxostoma pidiense Cope. Vse. (155)
181. Moxostoma coregonus Cope. Vse. (156)
182. Moxostoma album Cope. Vse. (157)
183. Moxostoma thalassinum Cope. Vse. (158)
184. Moxostoma valenciennesi[2] Jordan. Vn. (159)
185. Moxostoma macrolepidotum Le Sueur. Ve. (160)
185 b. *Moxostoma macrolepidotum duquesnei* Le Sueur. Vw.
186. Moxostoma aureolum[3] Le Sueur. Vn. (161)
187. Moxostoma crassilabre Cope. Vse. (162)
188. Moxostoma congestum[4] Cope. Vsw. (166)

---

terorbital space $4\frac{1}{2}$; air-bladder with two cells; D. 12; A. 1, 8; scales, 13–65–11. Pyramid Lake, Nevada; in deep water. (*Cope.*) (*Chasmistes cujus* Cope, Proc. Ac. Nat. Sci., Phila., 1883, 149.)

This paper "On the Fishes of the Recent and Pliocene Lakes of the Western Part of the Great Basin and of the Idaho Pliocene Lake" contains an important discussion of the fish fauna of Nevada, Oregon, and Idaho, with description of numerous fossil forms not long extinct and closely allied to recent *Cyprinidæ* and *Catostomidæ*.

[1] The two forms of *Erimyzon* described in the Synopsis as *E. sucetta* and *E. goodei* seem to be geographical varieties of one species, southern specimens having the scales considerably larger and more regularly arranged than in northern ones. To the southern form belong the typical examples of *Moxostoma kennerlyi* Girard and *Erimyzon goodei* Jordan. Specimens of this form have been examined by me, from streams of South Carolina, Georgia, Florida, Alabama, Louisiana, Illinois, and Texas. From Alabama, Louisiana, and Illinois I have seen specimens more or less distinctly intermediate, while from Virginia to Indian Territory (types *M. claviformis*) and northward only the small-scaled form occurs. It is probable that the original description of *Cat. sucetta* Lac. belongs to the southern form (*kennerlyi = goodei*). The northern form may then retain Mitchill's name, *oblongus*.

[2] *Moxostoma valenciennesi* Jordan, Proc. U. S. Nat. Mus., 1885 = *Catostomus carpio* C. & V., not of Raf.

[3] I now omit from the list, *Moxostoma bucco* Cope, based on the young of some species, probably of *M. aureolum*.

[4] I have recently found the types of *Catostomus congestus* and *Ptychostomus albidus*. They belong to the same species, a species shown by the late explorations of Jordan & Gilbert in Texas, to be very abundant in the waters of that State. The type of *P. albidus* has 44 scales in the lateral line instead of 56 as shown in Girard's figure. The specimens from Ash Creek, Arizona, referred with doubt to this species by Cope & Yarrow (Lieutenant Wheeler's Expl. Zoölogy, V. 680, 1876) belong apparently to *M. congestum*. The following account is taken from specimens taken by us in Lampasas River, at Belton, Tex.:

General form of *M. aureolum*, rather robust, moderately compressed, the back somewhat elevated. Head comparatively short, rather broad above and pointed anteriorly;

189. **Moxostoma conus** Cope. Vsc. (163)
190. **Moxostoma anisurum** Rafinesque. Vw. (164)
191. **Moxostoma pœcilurum** Jordan. Vsw. (165)
192. **Moxostoma cervinum** Cope. Vsc. (167)

66.—PLACOPHARYNX Cope. (69)

193. **Placopharynx carinatus** Cope.[1] Vw. (168)

67.—QUASSILABIA Jordan & Brayton. (70)

194. **Quassilabia lacera** Jordan & Brayton. Vw. (169)

Family XXXII.—CYPRINIDÆ. (31)

68.—CAMPOSTOMA Agassiz. (71)

195. **Campostoma ornatum**[2] Girard. Vsw. (170)
196. **Campostoma anomalum** Rafinesque. Vw. (171)
196b. *Campostoma anomalum prolixum* Storer. Ve. (172)
197. **Campostoma formosulum**[3] Girard. Vsw. (173)

69.—OXYGENEUM Forbes.

198. **Oxygeneum pulverulentum**[4] Forbes. Vw.

70.—ACROCHILUS Agassiz. (72)

199. **Acrochilus alutaceus** Agassiz & Pickering. T. (174)

71.—ORTHODON Girard. (73)

200. **Orthodon microlepidotus** Ayres. T. (175)

72.—LAVINIA Girard. (74)

201. **Lavinia exilicauda** Baird & Girard. T. (176)

73.—CHROSOMUS Rafinesque. (75)

202. **Chrosomus erythrogaster** Rafinesque. V. (177, 179)
203. **Chrosomus oreas**[5] Cope. Ve. (178)

74.—ZOPHENDUM Jordan. (76)

204. **Zophendum siderium** Cope. R. (180)
205. **Zophendum plumbeum** Girard. Vsw. (181)

---

the snout a little projecting, mouth rather small, the lower lip full, formed as in *M. aureolum*; eye small, about 5 in head; dorsal fin unusually low and small, little elevated in front, its first ray, when depressed, reaching about to the middle of the last ray; caudal not deeply forked, the lobes equal; lower fins moderate.

Smoky yellowish-brown above, yellowish-silvery below; lower fins whitish; none of the fins red in life; the membranes of the dorsal always dusky. Head 4¼ to 4¾; depth 4; D. 12; scales 6-45-5; teeth as in *M. aureolum*. Streams of Texas to Arizona.

[1] Professor Gilbert thinks that this species may be the original *Moxostoma anisurum* of Rafinesque.

[2] The types of *Campostoma ornatum* have 73 scales in the lateral line. Those of *C. nasutum* agree in all respects with the ordinary *C. anomalum*.

[3] The types of *Campostoma formosulum* have 46 scales in the lateral line.

[4] *Oxygeneum pulverulentum* Forbes, Bull. Ills. Lab. Nat. Hist., 1885, 136. Peoria, Ills.

[5] *Chrosomus oreas* is a doubtful species, which I have not yet examined. *C. eos* is doubtless indentical with *C. erythrogaster*.

75.—DIONDA[1] Girard. (77 pt.)

206. Dionda melanops Girard. Vsw. (189)
207. Dionda punctifera Garman. Vsw. (188b.)
208. Dionda fluviatilis Girard. Vsw. (188)
209. Dionda amara Girard. Vsw. (183)
210. Dionda episcopa[2] Girard. Vsw. (184, 187)
211. Dionda serena[3] Girard. Vsw. (185)
212. Dionda nubila[4] Forbes. Vw. (206)
213. Dionda (?) hæmatura[5] Cope. Vn. (204)

76.—HYBOGNATHUS Agassiz. (78)

214. Hybognathus meeki[6] Jordan & Gilbert. Vw.
215. Hybognathus argyritis[7] Girard. Vnw.
216. Hybognathus nuchalis[8] Agassiz. V. (182)
216 b. *Hybognathus nuchalis placita*[9] Girard. Vw. (186)

---

[1] The genus *Dionda* may perhaps be recognized as distinct from *Hybognathus*. Its teeth are shorter than those of *Hybognathus*, and more or less distinctly hooked. The species are small in size and mostly dusky in coloration, being especially characteristic of the Rio Grande region.

[2] *Dionda episcopa* Girard, *Dionda texensis* Girard, *Dionda argentosa* Girard (types of these three examined by us) = *Hybognathus flaripinnis* Cope. Fairly described in the Synopsis under the name of *Hybognathus flaripinnis*. The number of scales in the lateral line is about 37 in the types of *episcopa* and *argentosa*, 37 to 39 in *texensis*, and 41 in *flaripinnis*. The anterior suborbitals are of moderate width in *D. episcopa*, about as in *Hybognathus nuchalis*.

[3] *Dionda serena* Girard = *Dionda chrysitis* Grd. = *Hybognathus nigrotæniatus* Cope. Fairly described in the Synopsis under the latter name. The eye is smaller in *serena* than in *episcopa*, and the scales are larger (34 in the type of *D. serena*).

[4] Described in the Synopsis, page 167, as *Cliola nubila*. The species belongs, however, to *Dionda*, as has been already noticed by Professor Forbes. *D. nubila* is very close to *D. episcopa*, but from the specimens compared it appears to differ from the latter in the more pointed snout and in the larger mouth, the cleft of the mouth forming about one-fourth the length of the head, instead of one-fifth, as in *D. episcopa*.

[5] A doubtful species, unknown to me. The description points rather to this genus or *Cliola*, than to *Notropis*.

[6] *Hybognathus meeki* Jordan & Gilbert, Proc. U. S. Nat. Mus., 1885. Ozark region of Missouri and Arkansas: abundant.

[7] The types of *Hybognathus argyritis* from the Upper Missouri belong to a species distinct from *H. nuchalis*, and are distinct from the species heretofore called *H. argyritis* by different authors. The suborbitals in *H. argyritis* are broad, as in *H. nuchalis* and *H. placita*, the anterior being about twice as long as deep; the mouth is larger than in the other species, its cleft extending nearly to the eye; the jaws subequal, the lower being acutish at tip. The species is known only from the Upper Missouri and the Red River of the North. *Hybognathus evansi* Girard is possibly the same, but the types are lost and the description is too brief for identification. It is more likely *H. nuchalis*.

[8] This species ranges from New Jersey to South Carolina, Texas, and Dakota. *H. osmerinus* and *H. regius* being indistinguishable from it. It has the suborbitals broad, the mouth small, the lower jaw short, blunt, and subhorizontal, and the eye large, about 4 in head.

[9] *Hybognathus placita*, now known from the Arkansas and Missouri Rivers, is closely related to *H. nuchalis*, but has the eye smaller, about 5 in head, the snout depressed and rather blunt; mouth very small.

216c. *Hybognathus nuchalis regia* Girard. Vse.
217. **Hybognathus hayi**[1] Jordan. Vs. (182b.)

77.—PIMEPHALES[2] Rafinesque. (78, 79, 80)

218. **Pimephales promelas**[3] Rafinesque. V. (190, 191)
218b. *Pimephales promelas confertus* Girard. Vnw. (192)
219. **Pimephales notatus**[4] Rafinesque. V. (193, 194)

78.—EXOGLOSSUM Rafinesque. (81)

220. **Exoglossum maxillingua** Le Sueur. Ve. (195)

79.—COCHLOGNATHUS Baird & Girard. (82)

221. **Cochlognathus ornatus** Baird & Girard. Vsw. (196)
222. **Cochlognathus biguttatus** Cope. Vsw. (197)

80.—CLIOLA[5] Girard. (84 pt.)

223. **Cliola vigilax**[6] Baird & Girard. Vw. (202, 203, 215)

81.—NOTROPIS[7] Rafinesque. (83, 84, 85)

§ *Hemitremia.* (83)

224. **Notropis bifrenatus** Cope. Ve. (199)
225. **Notropis maculatus** Hay. Vs. (200)
226. **Notropis heterodon**[8] Cope. Vn. (201)

---

[1] *Hybognathus hayi* Jordan, Proc. U. S. Nat. Mus., 1884. Streams of Alabama, Mississippi, and the Lower Mississippi Valley. This species is correctly distinguished from *H. nuchalis* in the Synopsis, p. 968., under the erroneous name of *H. argyritis*. The species was first observed by Professor Hay.

[2] The genus *Hyborhynchus* is not distinct from *Pimephales*, the character of the lateral line being subject to many variations in *P. promelas*.

[3] *Coliscus parietalis* is, in my opinion, the young of *Pimephales promelas*. *Hyborhynchus confertus* is scarcely distinguishable from *P. promelas*, western specimens, Illinois to Texas, having the lateral line often complete, although usually more or less broken or irregular.

[4] *Hyborhynchus superciliosus* is not distinct from *Pimephales notatus*. The skin at the angle of the mouth is thickened and produced in the males, but there is no true barbel.

[5] CLIOLA Girard (type *Cliola vigilax*) = *Hypargyrus* Forbes, Proc. U. S. Nat. Mus., 1884, 200 (type *Hybopsis tuditanus* Cope), may be regarded as a genus distinct from *Notropis*, having the short intestines, curved teeth, and other characters of *Notropis*, with the separated first dorsal ray, and the general appearance of *Pimephales notatus*.

[6] *Cliola vigilax* B. & G. = *Cliola velox* Girard = *Cliola rivax* Girard = *Hybopsis tuditanus* Cope = *Alburnops taurocephalus* Hay. This widely-diffused and abundant species is described in detail by Professor Gilbert, Proc. U. S. Nat. Mus., 1884, 200, under the name of *Hypargyrus tuditanus*.

[7] I find it impossible to maintain the distinctions given in the Synopsis, of *Hemitremia*, *Cliola* and *Minnilus*. I therefore follow Professor Gilbert (Proc. U. S. Nat. Mus., 1884, 201) in uniting all these little fishes in a single genus, *Notropis*, the latter generic name being the earliest applied to any of the group.

[8] *Hemitremia vittata* is here omitted. The species is perhaps not distinct from *N. bifrenatus* or *N. heterodon*. In any case the name *vittatus* is preoccupied in *Notropis*. The number of teeth, 4-5, assigned to *H. vittata* by Professor Cope is probably an accidental variation or an error of observation. In some specimens, which as yet we are unable to separate from *N. heterodon*, the lateral line is complete, and the teeth 2, 4-4, 2. See Gilbert, Proc. U. S. Nat. Mus., 1884, 207.

§ *Alburnops* Girard.

227. **Notropis anogenus**[1] Forbes. Vw.
228. **Notropis spectrunculus** Cope. Vs. (205)
229. **Notropis illecebrosus**[2] Girard. Vw.
230. **Notropis ? fretensis**[3] Cope. Vn. (207)
231. **Notropis longirostris** Hay. Vs. (208)
232. **Notropis nitidus**[4] Girard. Vsw.
233. **Notropis deliciosus**[5] Girard. Vw. (213)
233 b. *Notropis deliciosus stramineus* Cope. Ve. (209)
233 c. *Notropis deliciosus longiceps* Cope. Ve. (211)
233 d. *Notropis deliciosus volucellus* Cope. Vn. (210)
234. **Notropis procne** Cope. Ve. (214)
235. **Notropis gilberti**[6] Jordan. Vw.

[1] *Notropis anogenus* Forbes. Bull. Ill. Lab. Nat. Hist., 1885, 138. Fox R., Ills.

[2] For description of this species see Proc. U. S. Nat. Mus., 1885. The original types of *N. illecebrosus* closely resemble those of *N. blennius*, differing especially in the form of the anterior suborbital which is in this species very narrow. The snout is less convex than in *N. blennius*. Abundant in Western Arkansas. We are unable to find Girard's type of *Alburnops shumardi*, and regard that species as doubtfully a synonym of *A. illecebrosus*.

[3] A doubtful species, unknown to me.

[4] *Moniana nitida* Girard, Proc. Ac. Nat. Sci., Phila., 1856, 201, erroneously referred, in the Synopsis (p. 175), to the synonymy of *Notropis deliciosus*. From the latter species Girard's types differ mainly in the larger, more oblique, and less inferior mouth. The following description is from the original type, from Cadereita, Nuevo Leon:

Head, $3\frac{2}{3}$; depth, $3\frac{3}{4}$; D. 8; A. 7; scales, 5-32-4. Body, stout, rather deep: eye, smallish, $3\frac{1}{4}$ in head; about equal to snout, and about $\frac{1}{2}$ less that interorbital area, which is quite flat; margin of upper lip on level with pupil; mouth rather large, oblique; snout little pointed; maxillary reaching slightly past vertical from front of orbit, its length about $3\frac{1}{3}$ in head; lower jaw shorter than upper, included when the mouth is closed; origin of dorsal slightly nearer tip of snout than base of caudal; about 12 scales in front of dorsal; tips of rays of dorsal all coterminous when the fin is deflexed; length of longest ray of dorsal $1\frac{1}{4}$ in head; base of fin scarcely 2 in head; anal similar to dorsal; longest, ray 2 in head; base, 3 in head; pectorals reaching $\frac{4}{5}$ distance to ventrals, $1\frac{2}{3}$ in head; ventrals reaching $\frac{3}{4}$ distance to anal, $1\frac{2}{3}$ in head; teeth, 4-4, little hooked; color, brownish, a faint silvery band along sides, little wider than diameter of eye, a very small faint dark spot at base of caudal; fins all plain. Two specimens from Cadereita.

[5] The types of *Moniana deliciosa* Girard, Proc. Acad. Nat. Sci. Phila., 1856, 199, are identical with the species described in the Synopsis as *Cliola missuriensis*. This form differs from *N. stramineus* Cope only in the somewhat greater size of the scales, there being 32 to 35 in the lateral line in *deliciosus*, 34 to 38 in *N. stramineus*. The latter, in our view, represents a slight variety found from Wisconsin to Tennessee, the true *deliciosus* ranging from Iowa to Texas.

Hybopsis longiceps Cope, from Virginia, appears also to represent a slight variety of *N. deliciosus*, with a more distinct dark lateral stripe, a rather longer preorbital region and slightly higher fins. Cope's type had the scales 5-33-2. A specimen from Fairfax, Va., has lat. l. 36. The identification of Rafinesque's *Minnilus microstomus* is too uncertain to warrant the use of his name.

Hybopsis volucellus Cope is unknown to me. It will probably prove to represent a variety of *N. deliciosus* with rather higher fins than usual.

[6] *Notropis gilberti* Jordan & Meek, Proc. U. S. Nat. Mus. 1884. It is abundant with *N. deliciosus* in the streams of Iowa, Kansas, and Missouri. From the latter it is readily distinguished by the smaller eye and soiled coloration.

236. **Notropis scylla** Cope. Vw. (212)
237. **Notropis nocomis** [1] Jordan & Gilbert. Vsw.
238. **Notropis phenacobius** [2] Forbes. Vw.
239. **Notropis chlorus** Jordan. Vnw. (216)
240. **Notropis comalis** [3] Jordan & Gilbert. Vsw.
241. **Notropis piptolepis** [4] Cope. (256)
242. **Notropis topeka** [5] Gilbert. V.
243. **Notropis boops** [6] Gilbert. V.
244. **Notropis blennius** [7] Girard. V. (275)
245. **Notropis simus** Cope. Vsw. (218)

§ *Hudsonius* Girard.

246. **Notropis hudsonius** [8] Clinton. Vnc. (221)
246 b. *Notropis hudsonius amarus* Girard. Vsc. (219. 220, 222)

§ *Codoma* Girard

247. **Notropis ornatus** Girard. Vsw. (226)

§ *Moniana* Girard.

248. **Notropis leoninus** [9] Girard. Vsw. (230)
249. **Notropis lutrensis** [10] Baird & Girard. Vw. (223, 224, 228, 229, 231, 238, 240)

---

[1] *Notropis nocomis* Jordan & Gilbert, Proc. U. S. Nat. Mus. 1885. Rio Comal, Texas.

[2] *Notropis phenacobius* Forbes, Bull. Ills. Lab. Nat. Hist., 1885, 137. Peoria, Ills.

[3] *Notropis comalis* Jordan & Gilbert, Proc. U. S. Nat. Mus., 1885. Rio Comal, Texas.

[4] *Photogenis piptolepis* Cope. Cope's description is repeated in the Synopsis, p. 183, under the erroneous name of *Cliola zonata* (Ag.). Agassiz's species is a very different one, allied to *N. coccogenis*.

[5] *Cliola topeka* Gilbert, Bull. Washburn, Lab. Nat. Hist. Kas., 1884, 1, 13; description reproduced, Proc. U. S. Nat. Mus., 1884. Western Iowa and Kansas. The male of this species is bright red in life.

[6] *Notropis boops* Gilbert, Proc. U. S. Nat. Mus., 1884, 201. Indiana to Missouri.

[7] *Alburnops blennius* Girard, Proc. Ac. Nat. Sci. Phila., 1856, 194. This species closely resembles *N. illecebrosus*, but its suborbital bones are very much broader than in the latter species, and its anterior profile is more decurved. One of Girard's types has the teeth 1, 4-4, 0. Arkansas River at Fort Smith.

[8] *Clupea hudsonia* Clinton, Ann. Lyc. N. H. N. Y., 1824 = *Hudsonius fluviatilis* Girard, Proc. Ac. Nat. Sci. Phila., 1856, 210 = *Luxilus selene* Jordan, Bull. U. S. Nat. Mus. X. 60, 1877. Great Lakes and streams eastward as far south as the Susquehanna. Southward (Maryland to Georgia) it is replaced by the subspecies *amarus*, which, as stated in the text, differs only in having the teeth 1, 4-4, 0 or 1, instead of 2, 4-4, 2 or 1, as in the typical *hudsonius*. *Alburnops saludanus* Jordan & Brayton, and *Hudsonius euryopa* Bean seem to be simply color variations of *amarus*. *Rutilus storerianus* Kirtland has been incorrectly identified with *N. amarus*, it being a species of *Hybopsis*, (= *Ceratichthys lucens* Jordan).

[9] *Moniana leonina, complanata*, and *frigida* Girard. Of these nominal species I have found the types of *M. frigida* only. These seem to represent a species distinct from *N. lutrensis*, having the caudal peduncle more elongate, and 37 scales in the lateral line.

[10] *Leuciscus lutrensis* Baird & Girard = *Hypsilepis iris* Cope = *Moniana jugalis* Cope = *Moniana gibbosa* Girard = *Cyprinella forbesi* Jordan = *Moniana pulchella* Girard = *Moniana couchi* Girard = *Moniana gracilis* Girard = *Moniana lætabilis* Grd. = *Moniana rutila* Grd. = *Cyprinella billingsiana* Cope = ? *Cyprinella suavis* Girard.

Examination of the original types of the above nominal species, and of thousands

250. **Notropis proserpina**[1] Girard. Vsw. (233)
251. **Notropis formosus** Girard. Vsw. (234)
252. **Notropis callisema** Jordan. Vse. (227)

§ *Cyprinella* Girard.

253. **Notropis bubalinus**[2] Baird & Girard. Vw. (235, 236, 337)
254. **Notropis lepidus** Girard. Vw. (239)
255. **Notropis ludibundus** Girard. Vw. (242)
256. **Notropis garmani**[3] Jordan. Vsw. (236b.)
257. **Notropis macrostomus** Girard. Vsw. (241)
258. **Notropis notatus**[4] Girard. Vsw. (243)
259. **Notropis venustus** Girard. Vsw. (244)
260. **Notropis cercostigma**[5] Cope. Vsw. (276)
260 b. *Notropis cercostigma stigmaturus* Jordan. Vs. (245, 253)
261. **Notropis whipplei**[6] Girard. Vn. (246, 247)
262. **Notropis galacturus** Cope. Vs. (248)
263. **Notropis camurus**[7] Jordan & Meek. Vw.
264. **Notropis eurystomus** Jordan. Vse. (249)
265. **Notropis niveus** Cope. Vse. (250)
266. **Notropis callistius** Jordan. Vs. (251)
267. **Notropis trichroistius** Jordan & Gilbert. Vs. (252)
268. **Notropis coeruleus** Jordan. Vs. (254)
269. **Notropis chloristius** Jordan & Brayton. Vse. (255)
270. **Notropis xænurus** Jordan. Vse. (257)
271. **Notropis pyrrhomelas** Cope. Vse. (258)
272. **Notropis hypselopterus** Günther. Vs. (259)

---

of specimens collected by the writer in different streams from Iowa to Southern Texas have convinced me that all belong to a single species, variable in depth of body according to sex and circumstances, but otherwise very constant.

[1] *Moniana proserpina* Girard, Proc. Ac. Nat. Sci. Phila., 1856, 199. This species is well separated from the others with which Dr. Girard has associated it, and seems to be the same as his *Moniana aurata*.

[2] *Leuciscus bubalinus* Baird & Girard = *Cyprinella umbrosa* Girard = *Cyprinella gunnisoni* Girard. The types of *C. umbrosa* have 32 scales in the lateral line; those of *C. gunnisoni* 34; the latter are young examples of the same species.

[3] *Cyprinella rubripinna* Garman, Bull. Mus. Comp. Zool., 1881, VIII, 91. The name rubripinna (*rubripinnis*) is twice preoccupied in the genus *Notropis*, as here understood.

[4] *Cyprinella notata* Girard. This is apparently a valid species, very close to *N. cercostigma*, but with larger scales (34) and a much fainter caudal spot. Specimens from Austin, Tex., agree fairly with Girard's types, which are in very bad condition.

[5] *Cyprinella cercostigma* Cope = *Luxilus chickasavensis* Hay = *Cliola urostigma* Jordan & Meek, Proc. U. S. Nat. Mus., 1884, 475. Specimens examined from Pearl River, Mississippi, and from nearly all the rivers of Texas from the Red to the Nueces. In all these specimens the number of scales in the lateral line is 37 to 39, while in specimens from the Alabama Basin (Etowah, Coosa, Alabama, Black Warrior) the number is from 42 to 44. I regard these as an Eastern variety, *stigmaturus* (*Photogenis stigmaturus* Jordan = *Cyprinella calliura* Jordan). Excepting the size of the scales and the more orange coloration of the fins in the var. *cercostigma*, I can detect no constant difference.

[6] I cannot distinguish *N. analostanus* from *N. whipplei*. Arkansas specimens have the body usually a little more elongate, but are not otherwise different.

[7] *Cliola camura* Jordan & Meek, Proc. U. S. Nat. Mus., 1884, 474. Arkansas Basin, Colorado to Missouri.

§ *Luxilus* Rafinesque.

273. **Notropis megalops** [1] Rafinesque. Vn. (260, 272)
273b. *Notropis megalops frontalis* Agassiz. Vn.
273c. *Notropis megalops cyaneus* Cope. Ve.
274. **Notropis coccogenis** Cope. Vse. (262)
275. **Notropis zonatus** [2] Agassiz. Vw.
276. **Notropis zonistius** Jordan. Vse. (263)

§ *Hydrophlox* [3] Jordan & Brayton.

277. **Notropis roseus** Jordan. Vs. (264)
278. **Notropis rubricroceus** Cope. Vse. (265)
279. **Notropis lutipinnis** Jordan & Brayton. Vse. (266)
280. **Notropis chlorocephalus** Cope. Vse. (267)
281. **Notropis chiliticus** Cope. Vse. (268)
282. **Notropis chalybæus** Cope. Ve. (269)
283. **Notropis chrosomus** Jordan. Vs. (270)
284. **Notropis xænocephalus** Jordan. Vs. (271)
285. **Notropis lacertosus** Cope. Vs. (273)
286. **Notropis ariommus** [1] Cope. Ve. (277)
287. **Notropis scabriceps** Cope. Vw. (278)
288. **Notropis jejunus** Forbes. Vw. (279)
289. **Notropis leuciodus** Cope. Vs. (280)
290. **Notropis spilurus** [6] Gilbert & Swain. Vs.
291. **Notropis altipinnis** Cope. Vs. (281)
292. **Notropis amabilis** Girard. Vsw. (282)
293. **Notropis socius** Girard. Vsw. (283)
294. **Notropis swaini** [d] Jordan & Gilbert. Vsw.
295. **Notropis ? bivittatus** Cope. Vw. (284)

§ *Lythrurus* Jordan.

296. **Notropis ardens** [7] Cope. Vs. (289)
296b. *Notropis ardens lythrurus* Jordan. Vn. (288)
296c. *Notropis ardens atripes* Jordan. Vw. (287)
296d. *Notropis ardens cyanocephalus* Copeland. Vn. (286)

---

[1] *Cyprinus megalops* Rafinesque, Amer. Monthly Magazine and Crit. Review, I, 121, December, 1817 = *Cyprinus cornutus* Mitchill, Amer. Monthly Mag., II, 324, February, 1818. The name of Rafinesque has, therefore, priority.

Hybopsis plumbeolus Cope seems to have been based on a young specimen of this species.

[2] *Alburnus zonatus* Agassiz, Bull. Mus. Comp. Zool., 1, 9, 1863. Abundant in the Ozark region of Missouri and Arkansas; a beautiful species, closely allied to *N. coccogenis*, but with smaller mouth and different coloration. For detailed description see Jordan & Gilbert, Proc. U. S. Nat. Mus., 1885.

[3] As the typical species of *Alburnops* Girard (*blennius*) has the teeth 1, 4-4, 0, the name *Hydrophlox* may be adopted for this section, while *Alburnops* should supersede *Miniellus*.

[4] *Notropis spilurus* Gilbert & Swain, Proc. U. S. Nat. Mus., 1885. Northern Alabama.

[5] *Alburnellus megalops* Girard. The name *megalops* is preoccupied in this genus. For a description of this abundant species, see Jordan, Proc. U. S. Nat. Mus., 1885.

[6] I now regard the forms called in the Synopsis, *diplæmius* (*Minnilus diplæmius* Auct. (not *Semotilus diplæmius* Rafinesque) = *Notropis lythrurus* Jordan, Proc. U. S. Nat. Mus., 1884, 476), *atripes*, *cyanocephalus*, and *ardens* as varieties of a single species, of which the oldest tenable specific name is that of *ardens* Cope.

[7] *Alburnellus umbratilis* Girard = *Minnilus nigripinnis* Gilbert, Bull. Washb. Lab. N. H., 1, 1884, 14 = *Luxilus lucidus* Girard = ? *Notropis macrolepidotus* Forbes. Bull. Ills. Lab. Nat. Hist., 1885. 138. Iowa to Arkansas, very abundant. See Jordan & Gilbert, Proc. U. S. Nat. Mus., 1885.

CATALOGUE OF THE FISHES OF NORTH AMERICA.

297. **Notropis umbratilis**[1] Girard. Vw. (296, 416)
298. **Notropis punctulatus** Hay. Vs. (290)
299. **Notropis roseipinnis**[2] Hay. Vs. (291)
300. **Notropis bellus** Hay. Vs. (292)
301. **Notropis matutinus** Cope. Vse. (293)
302. **Notropis lirus**[3] Jordan. Vs. (294)
303. **Notropis metallicus** Jordan & Meek. Vse.

§ *Notropis.*

304. **Notropis scepticus** Jordan & Gilbert. Vse. (297)
305. **Notropis photogenis** Cope. Vse. (298)
306. **Notropis telescopus** Cope. Vs. (299)
307. **Notropis stilbius** Jordan. Vs. (300)
308. **Notropis atherinoides**[4] Rafinesque. Vn. (302)
309. **Notropis dilectus**[5] Girard. Vw. (295, 303, 305)
310. **Notropis rubrifrons**[6] Cope. Vn. (301, 304)
311. **Notropis micropteryx** Cope. Vw. (306)

§ *Protoporus*[7] Cope. (86)

312. **Notropis ? domninus** Cope. R. (307)
313. **Notropis ? timpanogensis** Cope. R. (285)

82.—ERICYMBA Cope. (87)

314. **Ericymba buccata** Cope. Ve. (308)

83.—PHENACOBIUS Cope. (88)

315. **Phenacobius teretulus** Cope. Ve. (309)
316. **Phenacobius mirabilis** Girard. Vw. (310, 310b.)
317. **Phenacobius catastomus** Jordan. Vs. (311)
318. **Phenacobius uranops** Cope. Vs. (312)

84.—TIAROGA Girard.

319. **Tiaroga cobitis** Girard. R. (217)

85.—RHINICHTHYS Agassiz. (89)

320. **Rhinichthys cataractæ**[8] Cuv. & Val. Vu. (313)
320 b. *Rhinichthys cataractæ dulcis* Girard. Vw. (314)

---

[1] *Notropis roseipinnis* Hay, nom. sp. nov., for *Minnilus rubripinnis* Hay. The name *rubripinnis* is preoccupied in this genus. *Argyreus rubripinnis* Heckel = *Notropis megalops*.

[2] *Notropis alabamæ* Jordan & Meek, Proc. U. S. Nat. Mus., 1884, 476, seems to be identical with *Notropis lirus*, which again is doubtfully distinct from *N. matutinus*.

[3] *Notropis metallicus* Jordan & Meek, Proc. U. S. Nat. Mus., 1884, 475. Allamaha (Suwannee) River, Georgia.

[4] *Notropis atherinoides* Rafinesque = *Alburnus rubellus* Agassiz =? *Minnilus dinemus* Rafinesque. The synonymy of this and related species is at present in much confusion.

[5] The types of *Alburnellus jemezanus* are shriveled and distorted. I am unable to see how they differ from *N. dilectus*.

[6] *Alburnellus percobromus* Cope seems to be indistinguishable from *N. rubrifrons*.

[7] The genus *Protoporus* is extremely doubtful, both the species referred to it being probably the young of *Squalius* or *Phoxinus*.

[8] Examination of large numbers of specimens of *Rhinichthys* from various parts of the United States has convinced me that not more than two distinct species can be

320 c. *Rhinichthys cataractæ transmontanus* Cope. R. (315)
321. **Rhinichthys atronasus** Mitchill. Vn. (316, 317)

86.—AGOSIA Girard. (90)

§ *Agosia.*

322. **Agosia chrysogaster** Girard. R. (318)
323. **Agosia metallica** Girard. R. (319)
324. **Agosia novemradiata**[1] Cope. R.

§ *Apocope* Cope. (91)

325. **Agosia carringtoni** Cope. R. (320)
326. **Agosia nubila**[2] Girard. R. (321, 322, 323, 324)
327. **Agosia oscula**[3] Girard. R. (325)

87.— HYBOPSIS[4] Agassiz (92)

§ *Nocomis* Girard.

328. **Hybopsis biguttatus**[5] Kirtland. V. (325, 327)

§ *Hybopsis.*

329. **Hybopsis cumingi** Günther. T. ? (329)
330. **Hybopsis storerianus**[6] Kirtland. Vw. (330)

---

recognized. *R. transmontanus* represents a tangible variety, occurring west of the Rocky Mountains and having a greater number of scales below the lateral line than I have ever seen in *R. cataractæ*. *Rh. dulcis* has the snout shorter and blunter than usual in *cataractæ*, projecting little beyond the mouth. Garman's review of this genus (Science Observer, 1881, 57) seems to me worse than useless.

[1] *Agosia novemradiata* Cope, Proc. Ac. Nat. Sci. Phila., 1883, 141. Silvery, dusted with smoky above and marked on sides with several rows of dusky spots; bases of lower fins and upper lip red; head elongate, especially the muzzle, which projects a little; eye $4\frac{1}{2}$ in head, $1\frac{1}{2}$ in muzzle, and in interorbital width; dorsal inserted behind ventrals; caudal peduncle rather deep; head 4; depth 5; D. always 1, 9; A. 1, 7; scales 11–60–11. Weber River, at Echo, Utah. (*Cope.*)

[2] On comparison of many examples, including the original types of *Apocope nubila*, *vulnerata*, and *henshavii*, I am unable to appreciate any permanent specific distinctions. The genus *Apocope* is scarcely distinct from *Agosia*.

[3] *Argyreus osculus* Girard = *Argyreus notabilis* Girard = *Apocope ventricosa* Cope. This species differs from *A. nubila* chiefly in the much smaller size of the scales. The original type of *A. osculus* has 90 scales in the lateral line, which is nearly complete.

[4] There is little doubt of the identity of *Hybopsis gracilis* Agassiz with *Ceratichthys amblops*. The name *Hybopsis* is therefore prior both to *Nocomis* and *Ceratichthys* as the designation of this genus.

[5] *Ceratichthys micropogon* Cope is probably based on an abnormal individual of *H. biguttatus*.

[6] *Rutilus storerianus* Kirtland = *Ceratichthys lucens* Jordan. By a curious mistake, Kirtland's species has been confounded by several recent writers with *Notropis amarus*, a species similar in appearance but lacking barbels. This handsome species reaches a length of 10 inches and is abundant in the lakes and river channels of the Mississippi Valley and the lake region. The teeth are usually 1, 4–4, 0.

331. **Hybopsis amblops** Rafinesque. Vw. (331)
331b. *Hybopsis amblops rubrifrons* Jordan. Vse. (332)
332. **Hybopsis hypsinotus** Cope. Vse. (333)

§ *Erinemus* Jordan.

333. **Hybopsis dissimilis** Kirtland. Vn. (334)
334. **Hybopsis monachus** Cope. Vs. (340)
335. **Hybopsis zanemus** Jordan & Brayton. Vse. (339)
336. **Hybopsis labrosus** Cope. Vse. (338)
337. **Hybopsis hyostomus**[1] Gilbert. Vw.
338. **Hybopsis montanus**[2] Meek. Vw.
339. **Hybopsis marconis**[3] Jordan & Gilbert. Vsw.
340. **Hybopsis æstivalis**[4] Girard. Vsw. (335, 336)
341. **Hybopsis gelidus**[5] Girard. Vnw. (337)

88.—COUESIUS Jordan. (93)

342. **Couesius squamilentus** Cope. Vnw. (341)
343. **Couesius dissimilis**[6] Girard. Vnw. (342.)
344. **Couesius plumbeus**[7] Agassiz. Vn. (343)
345. **Couesius physignathus** Cope. Vnw. (344)

89.—PLATYGOBIO Gill.

346. **Platygobio gracilis**[8] Richardson. Vnw. (345, 346)

90.—SEMOTILUS Rafinesque. (95)

347. **Semotilus atromaculatus**[9] Mitchill. V. (347)
348. **Semotilus thoreauianus** Jordan. Vs. (348)
349. **Semotilus bullaris** Rafinesque. Vnc. (349)

---

[1] *Nocomis hyostomus* Gilbert, Proc. U. S. Nat. Mus, 1884, 203. Indiana, Iowa, to Tennessee; not rare in river channels.

[2] *Hybopsis montanus* Meek, Proc. U. S. Nat. Mus, 1884. Upper Missouri region.

[3] *Hybopsis marconis* Jordan & Gilbert, Proc. U. S. Nat. Mus., 1885. Rio San Marcos, Texas.

[4] *Gobio æstivalis* Girard = *Ceratichthys sterletus* Cope. This species is allied to *H. hyostomus*, but has a much smaller eye; 4 to 4½ in head.

[5] *Hybopsis gelidus* is very pale in color, nearly or quite immaculate. The lower lobe of the caudal is dusky; the eye is small, 4 in head; and the scales are smaller than in related species, there being 44 in the lateral line. The barbel in these small fishes (*H. gelidus; æstivalis; hyostomus; zanemus; montanus; marconis,*) is much more developed than in any other of the American *Cyprinidæ*.

[6] The description in the Synopsis, of *Couesius dissimilis* is somewhat confused with that of *C. plumbeus*.

From the latter species *C. dissimilis* differs in the larger scales (60 instead of 68), the more decurved lateral line, and the more robust body. Mouth oblique, subterminal, resembling that of *Semotilus*. It is thus far known only from the Upper Missouri region.

[7] *Gobio plumbeus* Agassiz = *Nocomis milneri* Jordan = *Ceratichthys prosthemius* Cope. Adirondack region, northwest to Manitoba.

[8] I am unable to distinguish *Platygobio pallidus*, by the description, from *Platygobio gracilis*.

[9] The original *Cyprinus corporalis* of Mitchill is *Semotilus bullaris*. This species must therefore stand as *Semotilus atromaculatus*.

91.—POGONICHTHYS Girard. (96, 97)

350. Pogonichthys macrolepidotus[1] Ayres. T. (350, 351)

92.—STYPODON Garman. (97b.)

351. Stypodon signifer Garman. R. (352)

93.—MYLOCHILUS Agassiz. (98)

352. Mylochilus caurinus Richardson. T. (353)

94.—MYLOPHARODON Ayres. (99)

353. Mylopharodon conocephalus Baird & Girard. T. (225)

95.—PTYCHOCHILUS Agassiz. (100)

354. Ptychochilus oregonensis Richardson. T. (355)
355. Ptychochilus rapax[2] Girard. T. (356)
356. Ptychochilus harfordi Jordan & Gilbert. T. (357)
357. Ptychochilus lucius Girard. T. (358)

96.—GILA Baird & Girard. (101)

358. Gila elegans Baird & Girard. R. (359)
359. Gila robusta Baird & Girard. R. (360)
360. Gila grahami Baird & Girard. R. (361)
361. Gila affinis Abbott. R. (362)
362. Gila gracilis Baird & Girard. R. (363)
363. Gila emorii Baird & Girard. R. (364)
364. Gila nacrea Cope. R. (365)
365. Gila seminuda Cope & Yarrow. R. (366)

97.—PHOXINUS[3] Agassiz. (102, 103)

§ *Clinostomus* Girard.

366. Phoxinus elongatus Kirtland. Vn. (367)
367. Phoxinus vandoisulus Cuv. & Val. Ve. (368)
368. Phoxinus estor Jordan & Brayton. Vs. (369)
369. Phoxinus funduloides Girard. Ve. (370)

§ *Tigoma* Girard.

370. Phoxinus hydrophlox Cope. R. (371)
371. Phoxinus tænia Cope. R. (372)
372. Phoxinus montanus Cope. R. (373)
373. Phoxinus humboldti Girard. R. (374)

---

[1] The type of *Pogonichthys* (*Symmetrurus*) *argyriosus* is a young specimen of *Pogonichthys macrolepidotus*.

[2] The chief character in which the single known example of *P. rapax* differs from *P. oregonensis* is in the small size of the scales before the dorsal fin, there being 49 in *P. rapax* and about 42 in *P. oregonensis*.

[3] The character of the imperfection of the lateral line, which alone distinguishes *Phoxinus* from *Squalius*, as understood in the Synopsis, is of such slight importance and subject to such variations that I think best to merge the two groups in one. The name *Phoxinus* seems to have priority.

[31] CATALOGUE OF THE FISHES OF NORTH AMERICA.

374. Phoxinus galtiæ [1] Cope. R.
375 Phoxinus cruoreus Jordan & Gilbert. R. (375)
376. Phoxinus ardesiacus Cope. R. (376)
377. Phoxinus pandora Cope. R. (377)
378. Phoxinus margaritus Cope. Ve. (378)
379. Phoxinus gula Cope. R. (379)
380. Phoxinus pulcher Girard. R. (380)
381. Phoxinus egregius Girard. R. (381)
382. Phoxinus lineatus Girard. R. (382)
383. Phoxinus gracilis Girard. R. (383)
384. Phoxinus conformis Girard. T. (384)
385. Phoxinus bicolor Girard. T. (385)
386. Phoxinus obesus Girard. R. (386)
387. Phoxinus purpureus Girard. R. (387)
388. Phoxinus pulchellus Baird & Girard. R. (388)
389. Phoxinus intermedius Girard. R. (389)
390. Phoxinus aliciæ Jouy. R. (390)
391. Phoxinus copei Jordan & Gilbert. R. (391)
392. Phoxinus niger Cope. R. (392)
393. Phoxinus conspersus Garman. R. (393)

§ *Siboma* Girard.

394. Phoxinus crassicauda [2] Baird & Girard. T. (394)

§ *Squalius* Bonaparte.

395. Phoxinus atrarius [3] Girard. R. (395, 397)
396. Phoxinus squamatus Gill. (396)
397. Phoxinus crassus Girard. T. (398)

§ *Cheonda* Girard.

398. Phoxinus cœruleus Girard. T. (399)
399. Phoxinus cooperi Girard. T. (400)
400. Phoxinus nigrescens [4] Girard. R. (401)
401. Phoxinus modestus Garman. R. (402)

§ *Phoxinus*. (103)

402. Phoxinus neogæus Cope. Vn. (403)
403. Phoxinus flammeus Jordan & Gilbert. Vs. (404)
404. Phoxinus milnerianus Cope. Vnw. (405)
405. Phoxinus phlegethontis Cope. R. (406)

---

[1] *Squalius galtiæ* Cope, Proc. Ac. Nat. Sci. Phila., 1883, 148. Olive above as far as a plumbeous band which extends from the operculum to base of caudal. Below this line, sides and belly silver, except a broad band of crimson from the gill opening to front of anal; side of head with a dusky band. Dorsal inserted a little behind front of ventrals; muzzle short; mouth oblique, without prominent chin, the end of the maxillary reaching a little beyond front of orbit. Interorbital region gently and regularly convex as wide as eye. Head, 4; depth, 4¼; eye, 3 in head; D. 1, 8; A (probably) 8, scales 12-60-5; teeth 1, 4-5, 1, without grinding surface. Pyramid Lake, Nevada; abundant. (*Cope*.)

[2] The earlier name, *Leuciscus gibbosus* Ayres, is preoccupied by *Leuciscus gibbosus* Storer.

[3] I have no doubt that *Squalius rhomaleus* Jordan & Gilbert is the adult form of *P. atrarius*. *P. squamatus* is, perhaps, also the same species. Several of the species of *Phoxinus* here admitted are of very doubtful validity.

[4] *Tigoma nigrescens* Girard = *Squalius lemmoni* Rosa Smith, Proc. Cal. Ac. Sci., 1883. *P. modestus* is perhaps also this species.

98.—ALGANSEA[1] Girard. (104)

406. Algansea obesa Girard. R. (408)
407. Algansea symmetrica[2] Baird & Girard. T. (409)
408. Algansea bicolor Girard. T. (410)
409. Algansea parovana[3] Cope. R. (411)
410. Algansea thalassina[4] Cope.
411. Algansea antica Cope. Vsw. (412)
412. Algansea olivacea[5] Cope. R.
413. Algansea dimidiata[6] Cope. R.

§ *Siphateles* Cope.
414. Algansea vittata[7] Cope. R.

---

[1] *Leucos* Heckel (preoccupied) = *Algansea* Girard = *Myloleucus* Cope. Professor Cope (Proc. Ac. Nat. Sci. Phila., 1883, 142) recognizes *Myloleucus* and *Leucus* as distinct genera; the former with teeth 4–5; the latter 5–5. Besides these, he proposes a third genus, *Siphateles* (l. c. 146), having the teeth 5–5, with grinding surface, and the lateral line incomplete. Such minute subdivision seems to me undesirable.

[2] *Pogonichthys symmetricus* Baird & Girard (Proc. Ac. Nat. Sci. Phila., 1854, 136) = *Algansea formosa* Girard (l. c. 1856, 183). The original type of *P. symmetricus* has the teeth 4–5, the maxillary without barbel, the head 4 in length, the depth 4¼. Scales 9-53-6. I cannot distinguish it from *Algansea formosa*.

[3] Professor Cope regards *Myloleucus parovanus* as distinct from *Algansea bicolor*. It is described as follows:

Translucent, with a plumbeous lateral band; ventrals and pectoral, dusky; dorsal and caudal shaded with dark; body, rather stout; muzzle, short, conical; mouth, very broad, the maxillary reaching front of orbit; profile, gently arched; eye, large, 3 in head, equal to interorbital width; pectorals reaching little more than half way to ventrals; the latter just to vent. Head, 3¼; depth, 4¼. D. 1,9; A.1,8. Scales, 10-48-5. Teeth, 4-5. L., 12 inches (*Cope*). Beaver River, Utah; Goose Lake and Klamath Lake, Oregon; abundant.

(*Myloleucus parovanus* Cope, Proc. Am. Phil. Soc. Phila., 1874, 136; Cope & Yarrow, Zoöl. Wheeler Sur, V. 669, 1876; Cope, Proc. Ac. Nat. Sci. Phila., 1883, 143.)

[4] *Myloleucus thalassinus* Cope. Slenderer than *M. parovanus*, and the color a light translucent green, quite unlike the heavy olivaceous of the latter. Head, 3¼; depth, 4¼. A. 1, 9. Scales, 9-46-4. Teeth, 4-5. L., 6 inches. One specimen known, from Goose Lake, Oregon. (*Cope*, Proc. Ac. Nat. Sci. Phila., 1883, 143.)

[5] *Leucus olivaceus* Cope. Dusky olive; the belly silvery; no lateral band; fins dusky; body fusiform, compressed; head narrowed to the muzzle, the mouth opening obliquely forwards and upwards; maxillary concealed in the closed mouth, its tip extending a little beyond front of eye. Eye 1¼ in snout, 1¾ in interorbital space, 5 in head, middle of front flat, its edges sloping to the superciliary border. Head, 3⅞; depth 4. A. 1, 8. Scales, 13-58-7. Teeth, 5-5, sharp edged. L., 1 foot. Pyramid Lake, Nevada; very abundant. (*Leucus olivaceus* Cope. Proc. Ac. Nat. Sci. Phila., 1883, 145.)

[6] *Leucus dimidiatus* Cope. Light brown above, becoming plumbeous lower, the belly pure silver-white. Eye equal to interorbital width, 3½ in head, a little more than length of muzzle. Mouth oblique, the maxillary reaching front of eye. Ventral a little behind front of dorsal. Head, 4; depth 4¼. A. 1, 8. Scales, 14-65-8. Teeth, 5-5. L., 4 inches. Pyramid Lake, Nevada; very abundant.

(*Leucus dimidiatus* Cope, Proc. Ac. Nat. Sci. Phila., 1883, 146.)

[7] *Siphateles vittatus* Cope. Brownish above, belly and sides silvery; a straight lateral band of lead-color interrupted at base of caudal by a vertical band of straw-yellow, which has a dark posterior edge. Lateral line very imperfect. Eye, 3 in head, a little less than interorbital width. Mouth oblique, the maxillary not quite reaching front of eye. Ventral fins beneath anterior part of dorsal. Head 4; depth,

99.—OPSOPŒODUS[1] Hay. (105, 106)

415. Opsopœodus emiliæ Hay. Vs. (413, 414)

100.—LUXILINUS[2] Jordan, (gen. nov.).

416. Luxilinus occidentalis Baird & Girard. T. (418)

101.—NOTEMIGONUS Rafinesque. (107)

417. Notemigonus gardoneus Cuv. & Val. Vsc. (415)
418. Notemigonus chrysoleucus[3] Mitchill. Vn. (417)
418 b. *Notemigonus chrysoleucus bosci* Cuv. & Val. Vsc. (419)

102.—RICHARDSONIUS Girard. (108)

419. Richardsonius balteatus Richardson. T. (421)
420. Richardsonius lateralis Girard. T. (422)

103.—LEPIDOMEDA Cope. (109)

421. Lepidoneda vittata Cope. R. (423)
422. Lepidomeda jarrovii Cope. R. (424)

104.—MEDA[4] Girard. (110, 111)

423. Meda fulgida Girard. R. (425)
424. Meda argentissima Cope. R. (426)

4½. D. 1, 8; A. 1, 8. Scales, 11-55-5. Teeth, 5-5, with well developed grinding surface. L., 3 inches. Pyramid Lake, Nevada. (Cope, Proc. Ac. Nat. Sci. Phila., 1883, 146.)

[1] The genus *Trycherodon* should be suppressed, its typical species, *T. megalops*, being identical with *Opsopœodus emiliæ*.

[2] LUXILINUS Jordan.

(Genus nova: type *Luxilus occidentalis* B. and G.) Ventral edge of moderate width; scaled over and not at all carinated; otherwise essentially as in *Notemigonus*. Gill rakers slender, of moderate length. Teeth 5-5 with entire edges and well developed grinding surface, their tips little hooked. Intestines of the short type, but longer than in most related genera. Anal basis elongate. (Name, a diminutive of *Luxilus*; from *lux*, light.)

[3] Specimens from Virginia, South Carolina, Georgia, and Florida (var. *bosci*) have 43 to 50 scales in the lateral line, and 15 to 17 rays in the anal fin. Specimens from various northern and western localities, Nova Scotia to Maryland, Louisiana, and Dakota (var. *chrysoleucus*) have 46 to 51 scales in the lateral line, and 12 to 14 anal rays. I regard the two forms as geographical varieties of one species. The name *Cyprinus americanus* is preoccupied, having been first given to a *Menticirrus*.

[4] The types of *Meda fulgida*, lately found by me, have the teeth 2, 5-5, 2, not 1,4-4, 1, as stated by Girard. The genus *Meda* is therefore identical with *Plagopterus*. The small barbel mentioned by Cope as a character of *Plagopterus*, I am unable to find either in *Meda* or *Plagopterus*.

Meda fulgida is closely allied to *Meda argentissima*, but has the eye a little larger, the snout shorter, the lower jaw more prominent. In form, size, coloration, and fin rays the two agree fully.

Family XXXIII.—CHARACINIDÆ. (32)

105.—TETRAGONOPTERUS Cuvier. (114)

§ *Astyanax* Baird & Girard.

425. Tetragonopterus argentatus Baird & Girard. Vsw. (429)

Order O.—ISOSPONDYLI. (M)

Family XXXIV.—ALEPOCEPHALIDÆ. (33)

106.—ALEPOCEPHALUS Risso. (115)

426. Alepocephalus bairdii Goode & Bean. B. (430)
427. Alepocephalus agassizii[1] Goode & Bean. B.
428. Alepocephalus productus[2] Gill. B.

Family XXXV.—ALBULIDÆ. (34)

107.—ALBULA (Gronow) Bloch & Schneider. (116)

429. Albula vulpes Linnæus. S. W. C. P. (116)

Family XXXVI.—HYODONTIDÆ. (35)

108.—HYODON Le Sueur. (117)

430. Hyodon alosoides Rafinesque. Vw. (432)
431. Hyodon tergisus Le Sueur. Vw. (433)
432. Hyodon selenops Jordan & Bean. Vsw. (434)

Family XXXVII.—ELOPIDÆ. (36)

109.—ELOPS Linnæus. (118)

433. Elops saurus Linnæus. S. W. P. (435)

110.—MEGALOPS Lacépède. (119)

434. Megalops atlanticus Cuv. & Val. S. W. (436)

---

[1] *Alepocephalus agassizii* Goode & Bean. Dusky; head and fins nearly black. Body a little deeper than in *A. bairdii*. Head compressed, the snout conically elongate, the lower jaw slightly produced; width of head 9½ in length of body (12 in *A. bairdii*). Eye 3½ in head (4½ in *A. bairdii*). Scales parchment-like. Dorsal inserted directly above vent, the distance from its origin to base of caudal one-third its distance from front of eye. Anal inserted under second ray of dorsal. Length of pectoral equal to diameter of eye and 10¾ in body. Ventral about one-sixth of head. Head 3; depth 5. D. 15; A. 17. Scales 10-90-11. Gulf Stream, lat. 30°, in 922 fathoms. (*Goode & Bean.*) (Goode & Bean, Bull. Mus. Comp. Zoöl, 1882, 215.)

[2] *Alepocephalus productus* Gill, Proc. U. S. Nat. Mus., 1883, 256. Gulf Stream, in deep water.

Family XXXVIII.—CHANIDÆ.[1]

111.—CHANOS[1] Lacépède.

435. Chanos chanos[1] Forskål. P.

Family XXXIX.—CLUPEIDÆ.

112.—DUSSUMIERIA[2] Cuvier & Valenciennes.

436. Dussumieria stolifera[3] Jordan & Gilbert. W.

113.—ETRUMEUS[4] Bleeker. (120)

437. Etrumeus teres DeKay. S. (437)

114.—CLUPEA Linnæus. (122, 123)

§ *Clupea.*

438. Clupea harengus Linnæus. G. N. Eu. (437)
439. Clupea mirabilis[5] Girard. A. C. (438, 440)

---

[1] Family CHANIDÆ.

Clupeoid fishes, with the body oblong, compressed, covered with small, firm, adherent scales. Lateral line distinct. Abdomen broad and flattish; snout depressed; mouth small, anterior, the lower jaw with a small symphyseal tubercle; no teeth. Premaxillary joined to upper anterior edge of maxillary. Gill membranes broadly united; free from the isthmus. Branchiostegals 4; pseudo-branchiæ well developed. An accessory branchial organ in a cavity behind the gill cavity. Dorsal fin opposite the ventrals; anal fin shorter than dorsal. Mucus membrane of œsophagus raised into a spiral fold. Intestine with many convolutions. Coloration silvery. Large fishes of the warmer parts of the Pacific. One genus and two species known (*Clupeidæ*; group *Chanina* Günther, VII, 473).

Genus CHANOS Lacépède.

(*Lutodeira* Kuhl.)

(Lacépède Hist. Nat. Poiss, V, 395, 1803; type *Mugil chanos* Forskål = *Chanos arabicus* Lacépède.) Characters of the genus included above. (Χανος, the open mouth.) *Chanos chanos* (Forskål). Pacific and Indian Oceans; abundant in the Gulf of California and southward to Panama.

(*Mugil chanos* Forskål Descr. Anim., 74; *Mugil salmoneus* Forster, Bloch & Schneider, 121; *Chanos salmoneus* Günther, VII, 473, and of recent authors generally.)

[2] DUSSUMIERIA Cuvier & Valenciennes.

(Hist. Nat. Poiss., XX, 467; type *Dussumieria acuta* Cuv. & Val.)

Body rather elongate, somewhat compressed; the abdomen rounded and without serratures. Mouth terminal, of moderate width, formed as in *Clupea*, but the maxillary more slender. Very small teeth in patches on jaws, palatines, pterygoids, and tongue. Scales cycloid, entire, very deciduous. Branchiostegals numerous, very slender. Ventrals inserted below middle or posterior part of dorsal; anal low, of moderate length. Pseudobranchiæ well developed; pyloric cœca numerous. (Dedicated to M. Dussumier, a correspondent of Valenciennes, and the original discoverer of the typical species.)

[3] *Dussumieria stolifera* Jordan & Gilbert, Proc. U. S. Nat. Mus., 1884, 25. Key West, Fla.

[4] The name *Etrumeus* is from *Etrumeiwasi*, the Japanese name of *Etrumeus micropus*. The genera, *Etrumeus* and *Spratelloides*, seem scarcely separable from *Dussumieria*.

[5] *Spratelloides bryoporus* Cope, the types of which species I have examined, seems to be identical with *Clupea mirabilis*.

§ *Sardinia* Poey.
440. **Clupea sagax** Jenyns. C. (441)
441. **Clupea pseudohispanica** Poey. W. (441b.)
§ *Pomolobus* Rafinesque.
442. **Clupea chrysochloris** Rafinesque. V. S. (442)
443. **Clupea mediocris** Mitchill. N. (443)
444. **Clupea vernalis** Mitchill. N. S. Am. (444)
445. **Clupea æstivalis** Mitchill. N. S. Am. (445)
§ *Alosa* Cuvier.
446. **Clupea sapidissima** Wilson. N. S. Am. (446)
§ *Harengula* Cuv. & Val. (123)
447. **Clupea sardina**[1] Poey. W.
448. **Clupea thrissina**[2] Jordan & Gilbert. P.
449. **Clupea pensacolæ** Goode & Bean. S. W. (447)
450. **Clupea stolifera**[3] Jordan & Gilbert. P.

115.—OPISTHONEMA[4] Gill. (124)

451. **Opisthonema oglinum**[5] Le Sueur. S. W. (448)

---

[1] *Clupea sardina* (Poey) *Sardina de ley,* "*Pilchard.*"
Greenish, sides silvery, the scales often shaded with light orange and dotted with black; a yellow scapular blotch; lips and dorsal fin yellow; older specimens with faint orange streaks along the rows of scales; tips of dorsal and caudal blackish. Body comparatively deep and compressed; lower jaw projecting; teeth in broad patches on jaws, vomer, palatines, and tongue; maxillary reaching nearly to middle of eye, 2¾ in head. Eye very large, considerably longer than snout, 2¼ in head; cheeks and opercles striate; gill rakers not very long, comparatively few; scales rather large, firm, each crossed by several conspicuous vertical ridges; scales not adherent, readily deciduous. Insertion of dorsal little before that of ventrals at a point considerably nearer snout than base of caudal. Dorsal a little higher than long, its free edge concave; anal low; pectorals nearly reaching ventrals, 1⅓ in head. Head, 3¼; depth, 3⅓; D. 1, 15; A. 18. Lat. l., 36. Ventral acutes about 15 + 10. L., 8 inches. Florida Keys to Cuba; abundant in schools. Readily distinguished from *Cl. pensacolæ* by the large eye and loose scales.

(*Harengula sardina* Poey, Memorias Cuba, II, 310, 1860; *Harengula sardina* Poey, Enum. Pisc. Cubens., 1875, 147; ?? *Clupea macrophthalma* Ranz., Nov. Com. Ac. Sci. Inst. Bonon., 1842, 320; ?? *Clupea humeralis* Cuv. & Val., XX, 293: not *Clupea macrophthalma* nor *Clupea humeralis* Günther. *Harengula sardina* Goode & Bean, Proc. U. S. Nat. Mus., 1879, 152; *Clupea sardina* Jordan, Proc. U. S. Nat. Mus., 1881, 106.)

[2] *Clupea thrissina* Jordan & Gilbert, Proc. U. S. Nat. Mus., 1882, 353. Capo San Lucas.

[3] *Clupea stolifera* Jordan & Gilbert, Proc. U. S. Nat. Mus., 1881, 339. Mazatlan to Panama.

[4] *Opisthonema oglinum* (Le Sueur) Goode & Bean.
Omit from the synonymy *Clupea thrissa*[5] Osbeck, and add:
(*Megalops oglina* and *M. notata* Le Sueur, Journ. Ac. Nat. Sci. Phila., I, 359, 361; *Chatoëssus signifer* DeKay, New York Fauna Fishes, 1842, 264: *Opisthonema oglinum* Goode & Bean MSS.)

[5] The original basis of *Clupea thrissa* L. was a fish brought by Lagerström from China and described by Linnæus's pupil, Odhel, in the Amœn. Academ., V, 251, as *Clupea thryza*. This is a species of *Dorosoma*. To this latter genus belongs also the *Clupea thrissa* of Osbeck. In the synonymy of *Clupea thrissa* of the tenth edition of the *Systema Naturæ*, several references to *Opisthonema* are included, while the *Clupea thrissa*, described in the twelfth edition as being received from Dr. Garden, is *Dorosoma cepedianum*. The *Clupea thrissa* of Broussonet and of most later authors is the *Opisthonema*, but the Linnæan name must go with the original intention of its author.

452. Opisthonema libertate[1] Günther. P.

116.—BREVOORTIA Gill. (125)

453. Brevoortia tyrannus Latrobe. N. S. (450)
453 b. *Brevoortia tyrannus patronus* Goode. S. (449)

117.—OPISTHOPTERUS[2] Gill.

454. Opisthopterus lutipinnis[3] Jordan & Gilbert. P.

Family XL.—DOROSOMIDÆ. (38)

118.—DOROSOMA Rafinesque. (126)

455. Dorosoma cepedianum Le Sueur. V. S. N. (451)
456. Dorosoma mexicanum Günther. S. (451 b)

Family XLI.—ENGRAULIDÆ. (39)

119.—STOLEPHORUS Lacépède. (127)

457. Stolephorus ringens Jenyns. C. P. (452)
458. Stolephorus macrolepidotus[4] Kner & Steindachner. P.
459. Stolephorus opercularis[5] Jordan & Gilbert. P.
460. Stolephorus browni Gmelin. N. S. W. (453)
461. Stolephorus perthecatus[6] Goode & Bean. S.

---

[1] *Meletta libertatis* Günther, Proc. Zool. Soc., Lond., 1866, 603; *Clupea libertatis* Günther, VII, 433; *Opisthonema libertate* Jordan & Gilbert, Proc. U. S. Nat. Mus., 1882, 622; Mazatlan to Panama, abundant.

[2] OPISTHOPTERUS Gill.

(Proc. Ac. Nat. Sci. Phil., 1861; 31; type *Pristigaster tartoor* Cuv. & Val.)
Body elongate, very much compressed, with the abdomen prominent and strongly serrated. Scales thin, deciduous, of moderate size. Lower jaw projecting: teeth rather small, in villiform bands on both jaws, palatines, pterygoids and tongue; vomer toothless. Dorsal fin small, considerably behind middle of body. Anal fin very long. Ventrals wanting. Caudal deeply forked. Tropical parts of the Pacific. (Ὄπισθη, behind: πτέρον, fin, the dorsal being placed farther backward than in the closely related genus *Pristigaster*.)

[3] *Pristigaster lutipinnis* Jordan & Gilbert, Proc. U. S. Nat. Mus., 1881, 340. Gulf of California and southward.

[4] *Stolephorus macrolepidotus* Kner & Steindachner. Body comparatively short and deep. Head one-fourth longer than deep. Snout very short, not projecting far beyond lower jaw. Jaws toothless. Maxillary narrow, rounded behind, extending to angle of preopercle. Abdomen slightly compressed. Scales adherent. Origin of dorsal slightly behind middle of body. Silvery, sides with an indistinct bluish band. Head 3½; depth 3, D. 12, A. 28. Scales 35-9. Mazatlan to Panama, one of the largest of the American species of *Stolephorus*.

(*Engraulis macrolepidotus* Kner & Steindachner, Abhandl. Bayer. Akad. Wiss. X, 1864; *Engraulis macrolepidotus* Günther, VII, 385.)

[5] *Stolephorus opercularis* Jordan &' Gilbert. Proc. U. S. Nat. Mus., 1881, 275. (Gulf of California.)

[6] *Stolephorus perthecatus* Goode & Bean., Proc. U. S. Nat. Mus., 1882, 434.
Pensacola, Fla. Apparently distinguished from *S. browni* by the short anal and from *S. perfasciatus* by the long maxillary.

462. Stolephorus ischanus[1] Jordan & Gilbert. P.
463. Stolephorus perfasciatus[2] Poey. W.
464. Stolephorus eurystole[3] Swain & Meek. N. (455)
465. Stolephorus curtus[4] Jordan & Gilbert. P.
466. Stolephorus mitchilli Cuv. & Val. N. S. (454 b.)
467. Stolephorus exiguus[5] Jordan & Gilbert. P.
468. Stolephorus miarchus[6] Jordan & Gilbert. W. P.
469. Stolephorus delicatissimus Girard. C.
470. Stolephorus lucidus[7] Jordan & Gilbert. P.
471. Stolephorus compressus Girard. C.

Family XLII.—ALEPIDOSAURIDÆ. (40)

120.—PLAGYODUS[8] Steller. (128)

472. Plagyodus ferox Lowe. B. (458)
473. Plagyodus æsculapius Bean. A. (458 b.)
474. Plagyodus borealis Gill. C. A. (459)

Family XLIII.—PARALEPIDIDÆ. (41)

121.—SUDIS Rafinesque. (129)

§ *Sudis.*

475. Sudis ringens Jordan & Gilbert. B. P. (459)

§ *Arctozenus* Gill.

476. Sudis borealis[9] Reinhardt. G. A. B. (461, 462)

---

[1] *Stolephorus ischanus* Jordan & Gilbert., Proc. U. S. Nat. Mus., 1881, 340. Mazatlan southward. Closely related to *S. browni*.

[2] *Stolephorus perfasciatus* (Poey).

Body rather elongate; snout compressed and pointed, shorter than eye. Top of head with a slight keel. Eye 3½ in head. Maxillary and lower jaw finely toothed; maxillary unusually short, its posterior end rounded, not extending quite to margin of preopercle; gill rakers numerous; pectoral 1⅔ in head, not reaching ventrals; insertion of anal below last rays of dorsal, the fin short; origin of dorsal midway between root of caudal and pupil. Color of *S. browni*, the lateral band rather narrower, well defined, its width about ⅔ eye; no dark punctulations except on base of caudal and sometimes on anal. Head 4¼; depth 6, D. 12, A. 14 to 16, L. 2 to 3 inches. (*Swain & Meek.*) Florida Keys to Cuba, common, but much less abundant than *S. browni*.

(*Engraulis perfasciatus* Poey, Mem. Cuba, II, 313, 1858. *Engraulis perfasciatus* Günther, VII, 391; not of Swain. Bull. U. S. Fish. Comm., 1882, 55, nor of Jor. & Gilb., Synopsis, 273; Swain & Meek, Proc. Ac. Nat. Sci. Phila. 1884.)

[3] *Stolephorus eurystole* Swain & Meek, Proc. Ac. Nat. Sci. Phila. 1884, 35. Wood's Holl, Mass. This is the species described in the Synopsis, p. 273, under the erroneous name of *S. perfasciatus*.

[4] *Stolephorus curtus* Jordan & Gilbert. Proc. U. S. Nat. Mus., 1881, 343. Mazatlan.

[5] *Stolephorus exiguus* Jordan & Gilbert. Proc. U. S. Nat. Mus., 1881, 342.

[6] *Stolephorus miarchus* Jordan & Gilbert. Proc. U. S. Nat. Mus., 1881, 344; 1882, 622; 1884, 106, Key West; Mazatlan, Panama. The smallest of the American anchovies.

[7] *Stolephorus lucidus* Jordan & Gilbert. Proc. U. S. Nat. Mus., 1881, 341. Mazatlan.

[8] It is probably best to substitute Steller's name, *Plagyodus*, for the later *Alepidosaurus*.

[9] *Sudis coruscans* is probably not specifically distinct from *S. borealis*.

Family XLIV.—SYNODONTIDÆ.[1] (42 part.)

122.—SYNODUS (Gronow) Bloch & Schneider.

§ *Synodus.*

477. **Synodus fœtens** Linnæus. S. (463)
478. **Synodus spixianus**[2] Poey. W.
479. **Synodus scituliceps**[3] Jordan & Gilbert. P.
480. **Synodus lucioceps** Ayres. C. (464)
481. **Synodus anolis**[4] Cuv. & Val. W. (464b.)

§ *Trachinocephalus* Gill.

482. **Synodus myops** Forster. S. W. (465)

123.—BATHYSAURUS[5] Günther.

483. **Bathysaurus agassizii** Goode & Bean. B.

Family XLV.—SCOPELIDÆ. (42)

124.—MYCTOPHUM Rafinesque. (131)

484. **Myctophum crenulare** Jordan & Gilbert. C. (466)

---

[1] Apparently those genera of the group called in the synopsis *Scopelidæ*, which have the maxillary rudimentary and adnate to the premaxillary, or sometimes entirely wanting, should be detached from *Scopelidæ*, to form a separate family, which has been called *Synodontidæ* by Professor Gill. To this group belong, in our fauna, the genera *Synodus* and *Bathysaurus*, as well as the Old World genera of *Harpodon* and *Saurida*.

[2] *Synodus spixianus* Poey. *Lagarto : Soap-fish.*

Sandy gray, light or dark, much mottled above with darker olive; branchiostegals pale yellowish; top of head without distinct vermiculations; dorsal scarcely barred; caudal dusky; other fins pale, with little or no yellow in life; lower parts of head mottled with dusky. No scapular spot; tip of snout not black. General form and appearance of *S. fœtens*, the teeth rather stronger; the jaws a little longer; the upper 1¼ in head. Dorsal fin shorter and higher, its free edge more oblique than in *S. fœtens*, its anterior rays when depressed extending beyond the tips of the posterior, 1¾ in head. Scales about as in *S. fœtens*. Pectorals 2 in head; ventrals 1¼. D. 1, 9. A. 11 or 12. Lat. l. 60. Florida Keys and Cuba. Abundant.

(*Saurus spixianus* Poey. Memorias Cuba, ii, 304, 1860; Poey, Enum. Pisc. Cubens., 1875, 141, Jordan, Proc. U. S. Nat. Mus., 1884, 107.)

For a detailed account of this and other American species of *Synodus*, see Meek, Proc. Ac. Nat. Sci. Phila., 1884, 130.

[3] *Synodus scituliceps* Jordan & Gilbert, Proc. U. S. Nat. Mus., 1881, 344. Mazatlan to Panama.

[4] The species described in the Synopsis (p. 889) as *Synodus intermedius*, is not that species, but a different one, *Saurus anolis* Cuv. & Val., xxii, 1849, 438 = *Synodus cubanus* Poey, Enum. Pisc. Cubens. 1875, 143. *Saurus intermedius* Agassiz & Spix. = *Synodus intermedius* Poey, Enum. Pisc. Cubens. 1875, 143, has the mouth smaller than in *S. anolis*, the scales larger (lat. l. 45), the scapular region without distinct black spot, and the coloration less variegated. *S. intermedius* is common in Cuba, but has not yet been noticed in our waters. In the adult of *S. anolis*, the lower parts are marked by stripes formed by an orange spot on each scale; the number of cross-bars is usually doubled by the presence of a shorter one between each pair.

[5] BATHYSAURUS Günther.

(Günther Ann. Mag. Nat. Hist., Aug., 1878, 181); type *Bathysaurus ferox* Günther.) Body formed as in *Synodus*, subcylindrical, elongate, covered with small scales.

485. **Myctophum mülleri**[1] Gmelin. G. (467)
486. **Myctophum boops**[2] Richardson. A.

125.—MAUROLICUS[3] Cocco. (132)

487. **Maurolicus borealis** Nilsson. B. (468)

Head depressed, with the snout produced, flat above. Cleft of the mouth very wide, with the lower jaws projecting; premaxillary very long, styliform, tapering, not movable; maxillary obsolete. Teeth in the jaws in broad bands, not covered by lips, curved, unequal in size, and barbed at the end; a series of similar teeth along the whole length of each side of the palate; a few teeth on the tongue, and groups of small teeth on the hyoid; eye moderate, lateral. Pectoral moderate; ventrals 8-rayed, inserted close behind pectoral. Dorsal fin median, of about 18 rays; adipose fin present or absent; anal moderate; caudal emarginate. Gill openings very wide, the gill membranes separate, free from the isthmus. Branchiostegals 11 or 12. Gill laminæ well developed; gill-rakers tubercular; pseudobranchiæ well developed. Scales rather small. Deep-sea fishes. (Βαθύς, deep; σαυρος, saurus = Synodus.)

Bathysaurus agassizii Goode & Bean.

Body elongate, subterete. Head alligator-like, naked, except on cheek and occiput, with strong nasal and interorbital ridges; its greatest width more than half its length; gape of mouth very wide, one-sixth length of body, extending behind eye for a distance equal to interorbital width. Premaxillary with two irregular rows of depressible teeth, some of them barbed, those of inner row much the largest; lower jaw enormously strong, its sides projecting beyond the upper jaw; its dentary edge thickly studded with depressible teeth, many of them, especially the larger inner ones, strongly barbed; those in front, claw-like, recurved; three rows of teeth on the palatines, the middle ones very much enlarged and most of them strongly barbed, these being the largest of all the teeth. On the tongue a few weaker teeth, and groups of similar teeth on the vomer. Insertion of dorsal behind snout at a distance a little more than its own base and about one-third the total length; longest ray equal to greatest depth of body. No adipose dorsal (in the specimen known); anal inserted considerably behind last ray of dorsal, its base about half that of the dorsal. Ventrals well apart, inserted just in front of dorsal, their length half head. Pectoral as long as lower jaw, its seventh ray prolonged to a length equal to that of head. Caudal slightly forked; scales thin, cycloid, deciduous, those of the lateral line larger, brownish; lining of gill cavity blue-black. Head, $3\frac{1}{4}$; depth, 7. B. 10, D. 17, A. 11, C. 19, P. 15, A. 8. Scales, 8–78–*. Length, 18 inches.

Gulf Stream, lat. 33°, at a depth of 647 fathoms. (*Goode & Bean.*)
(Goode & Bean, Bull. Mus. Comp. Zoöl., 1882, 215.)

[1] This species should stand as *Myctophum mülleri* instead of *M. glaciale*. To the synonymy add: *Salmo mülleri* Gmelin, Syst. Nat. 1788, 1378; *Scopelus mülleri*, Collet, Norske Nordhavs Exped., 1880, Fiske, 158; *Scopelus mülleri* Goode & Bean, Bull. Mus. Comp. Zoöl., 1882, 223.

This species has been lately taken in the deep waters off Southern New England.

[2] *Myctophum boops* Richardson.

Depth of head $1\frac{7}{8}$ in its length; eye nearly 3 in head; twice its distance from preopercle. Snout short, obtuse, its upper profile descending in a strong curve; jaws equal; maxillary reaching nearly to angle of preopercle, slightly and gradually dilated behind; cleft of mouth very slightly oblique. Origin of dorsal considerably nearer tip of snout than root of caudal, above base of ventrals; its last ray before origin of anal; pectoral reaching vent. Scales smooth, thin, and deciduous. Head $3\frac{1}{4}$; depth 5. D. 14. A. 21, V. 8 Scales 3–38–5. L. $4\frac{1}{2}$ inches. Vancouver's Island. (*Günther*).

(Richardson, Zoöl. Erebus and Terror. Fishes, 39, pl. 27. *Scopelus boops*, Günther, V, 408.)

[3] According to Professor Gill, the genus *Maurolicus* belongs to the *Scopelidæ* and not to the *Sternoptychidæ*.

Family XLVI.—HALOSAURIDÆ.[1]

126.—HALOSAURUS Günther.

488. **Halosaurus macrochir** Günther. B.

Family XLVII.—STOMIATIDÆ. (45)

127.—STOMIAS Cuvier. (134)

489. **Stomias ferox** Reinhardt. B. (470)

128.—HYPERCHORISTUS[2] Gill.

490. **Hyperchoristus tanneri** Gill. B.

---

[1] Family HALOSAURIDÆ.

Body elongate, compressed posteriorly, tapering into a very long and slender tail, which becomes compressed and narrowed into a sort of filament. Abdomen rounded. Scales rather small, cycloid, deciduous. Sides of head scaly; lateral line present, running along the sides of the belly, its scales, in the known species, enlarged, each in a pouch of black skin with a phosphorescent organ at its base. No barbels. Head subconical, depressed anteriorly, the flattened snout projecting beyond the mouth. Mouth inferior, horizontal, of moderate size, its anterior margin formed by the premaxillaries, its lateral margin by the maxillaries, which are of moderate width. Teeth small, in villiform bands, on the jaws, vomer, palatines, and tongue. Eye rather large. Facial bones with large muciferous cavities. Preopercle produced behind in a large flat process, "replacing the sub- and interoperculum." Bones of head unarmed. Gills 4, a slit behind the fourth. Pseudobranchiæ none. Gill-rakers short. Gill membranes separate, free from the isthmus. Branchiostegals numerous (about 14). Dorsal fin short, rather high, inserted behind ventrals and before vent. No adipose fin; no caudal fin. Anal fin extremely long, extending from the vent to the tip of the tail (its rays about 200 in number). Ventrals moderate, not very far back. Pectorals rather long, narrow, inserted high. No axillary scales. Air bladder large, simple. Stomach cœcal; pyloric cœca in moderate number; intestines short. Ovaries closed. No phosphorescent spots. A single genus, with about 5 species; fishes of the deep sea. (*Halosauridæ* Günther, VII, 482.)

HALOSAURUS Johnson.

(Johnson, Proc. Zoöl. Soc. London, 1863, 406; type *Halosaurus oweni*, Johnson, from Madeira). Characters of the genus included above. (ἅλς, sea; σαυρος, lizard.)

Halosaurus macrochir Günther.

Everywhere blackish, the color nearly uniform. Snout moderate, its length from mouth 7 in length of head; eye small, 7¼ in head, 2 in interorbital space. Length of head slightly greater than its distance from ventral. Maxillary reaching vertical from front of eye; its length from tip of snout 2⅜ in head. Insertion of dorsal entirely behind the ventrals. Ventrals midway between preopercle and front of anal, their length 2⅜ in head. Pectorals nearly reaching ventrals, 1¼ in head. Base of dorsal 3¼ in head, its longest ray 2. B. 12. D 1, 10, or 11, V. 9. Deep waters of the Atlantic; not rare in the Gulf Stream.

(Günther, Ann. Mag. Nat. Hist., 1878, 251; Goode & Bean, Bull. Mus. Comp. Zoöl., 1882, 219. *Halosaurus goodei* Gill, Proc. U. S. Nat. Mus., 1883, 257.)

[2] HYPERCHORISTUS Gill.

(Gill, Proc. U. S. Nat. Mus., 1883, 256; type, *Hyperchoristus tanneri* Gill.)

"Stomiatids, with a robust claviform body, naked skin, teeth on the jaws nearly uniserial, but in several groups, of which the successive teeth (about 4) rapidly

129.—ECHIOSTOMA Lowe. (135)

491. Echiostoma barbatum Lowe. B. (471)

130.—MALACOSTEUS[1] Ayres. (136)

492. Malacosteus niger Ayres. B. (472)

131.—ASTRONESTHES Richardson. (137)

493. Astronesthes niger Richardson. B. (473)

Family XLVIII.—ARGENTINIDÆ.[2] (46 part.)

132.—MICROSTOMA Cuvier. (138)

494. Microstoma grœnlandicum Reinhardt. G. (474)

133.—MALLOTUS Cuvier. (140)

495. Mallotus villosus Müller. A. G. (475, 476)

134.—THALEICHTHYS Girard. (141)

496. Thaleichthys pacificus Richardson. A. Ana. (477)

135.—OSMERUS Linnæus. (142)

497. Osmerus thaleichthys[3] Ayres. C. (478)
498. Osmerus mordax Mitchill. N. Ana. (480)
499. Osmerus dentex Steindachner. A. (481)

136.—HYPOMESUS Gill. (143)

500. Hypomesus pretiosus Girard. C. (482)
501. Hypomesus olidus Pallas. A. (483)

137.—ARGENTINA Linnæus.

502. Argentina syrtensium Goode & Bean. B. (484)

138.—HYPHALONEDRUS[4] Goode. (145)

503. Hyphalonedrus chalybeius Goode. B. (485)

increase in size backwards, and teeth on the palate enlarged, one on each side of the vomer and several on the palatines; moderate dorsals obliquely opposed, forked caudal and pectorals, each with a separate and specialized uppermost ray." ("Υπηρ, above; χοριστος, split, in allusion to the division of the pectorals.)

The species *H. tanneri* Gill, from the Gulf Stream in deep water, has not been described.

[1] According to Dr. Bean, the so-called barbel at the throat in *Malacosteus niger* is a muscle apparently concerned in the movement of the mandible.

[2] The *Argentininæ* may well be regarded as a family distinct from the *Salmonidæ*, differing in the form of the stomach, as stated in the Synopsis.

[3] *Osmerus attenuatus* Lockington, an extremely doubtful species, is here omitted, as also the land-locked varieties of *O. mordax*.

[4] This genus perhaps belongs to the *Scopelidæ*.

Family XLIX.—SALMONIDÆ. (46)

139.—COREGONUS Linnæus. (146)

§ *Prosopium* Milner.

504. Coregonus williamsoni Girard. R. (487)
505. Coregonus quadrilateralis Richardson. Vn. (488)
506. Coregonus kennicotti Milner. Y. (489)
507. Coregonus nelsoni[1] Bean. Y.

§ *Coregonus*.

508. Coregonus clupeiformis Mitchill. Vn. (490)
509 Coregonus labradoricus Richardson. Vn. (491)

§ *Argyrosomus* Agassiz.

510. Coregonus hoyi Gill. Vn. (492)
511. Coregonus merki Günther. Y. (493)
512. Coregonus laurettæ Bean. Y. (493 b.)
513. Coregonus artedi Le Sueur. Vn. (494)
514. Coregonus nigripinnis Gill. Vn. (495)

§ *Allosomus* Jordan.

515. Coregonus tullibee Richardson. Vn. (496)

140.—THYMALLUS Cuvier. (147)

516. Thymallus signifer Richardson. Y. Vn. (497)
516 b. *Thymallus signifer ontariensis*[2] Cuv. & Val. Vn. (497 b.)

141.—STENODUS[3] Richardson. (148)

517. Stenodus mackenziei Richardson. Y. Vn. (498)

142.—ONCORHYNCHUS Suckley. (149)

518. Oncorhynchus gorbuscha Walbaum. C. A. Ana. (499)

---

[1] *Coregonus nelsoni* Bean, Proc. U. S. Nat. Mus., 1884; waters of Alaska.

[2] *Thymallus ontariensis* Cuvier & Valenciennes, XXI, 452, 1848 (specimens sent by Milbert from Lake Ontario)=*Thymallus tricolor* Cope. The following is a translation of Valenciennes' account: We have received from Lake Ontario a *Thymallus* very near to that of the lake of Geneva. It has, however, more naked space under the throat, although less than in *Thymallus gymnothorax*. The head is evidently more pointed, the body more elongate, the dorsal a little longer. The denticulations of the scales are more pronounced. The colors seem scarcely to differ from those of *Thymallus*, for our specimens are greenish, with a dozen gray lines along the flanks. The dorsal has 4 or 5 longitudinal streaks of red. Our specimens are a foot long; they have been sent by M. Milbert. (*Valenciennes l. c.*)

[3] The original diagnosis of *Stenodus* is said to be in "Appendix Bach's Voyage. Rept. N. Am. Zoöl., 1836."

According to Dr. Bean, our species is probably not distinct from the Asiatic species, *S. leucichthys* (Guldenstadt).

519. **Oncorhynchus keta** Walbaum. C. A. Ana. (500)
520. **Oncorhynchus tchawytcha** Walbaum. C. A. Ana. (501)
521. **Oncorhynchus kisutch** Walbaum. C. A. Ana. (502)
522. **Oncorhynchus nerka** Walbaum. C. A. Ana. (503)

143.—SALMO Linnæus. (150)

§ *Salmo.*

523. **Salmo salar** L. N. En. Ana. (504)
523 b *Salmo salar sebago* Girard. Vne.

§ *Salar*[1] Cuv. & Val.

524. **Salmo gairdneri** Richardson. C. A. (506)
524 b *Salmo gairdneri irideus*[2] Ayres. T. (505)
525. **Salmo purpuratus** Pallas R. C. A. (508)
525 b. *Salmo purpuratus bouvieri* Bendire. R.
525 c. *Salmo purpuratus stomias* Cope. R.
525 d. *Salmo purpuratus henshawi* Gill & Jordan. R.
525 e. *Salmo purpuratus spilurus* Cope. R. (507)

144.—SALVELINUS Richardson. (151)

§ *Cristivomer* Gill & Jordan.

526. **Salvelinus namaycush** Walbaum. Vn. (509)
526 b *Salvelinus namaycush siscowet* Agassiz. Vn.

§ *Salvelinus.*

527. **Salvelinus oquassa**[3] Girard. Vne. (510, 511, 516?)
528. **Salvelinus arcturus** Günther. Vne. (512)
529. **Salvelinus malma** Walbaum. Y. C. A. (513)
530. **Salvelinus fontinalis** Mitchill. Vne. (514, 515)
530 b *Salvelinus fontinalis immaculatus* H. R. Storer. N. Ana.
531. **Salvelinus stagnalis**[4] Fabricius. G. (517?, 518)

Family L.—PERCOPSIDÆ.

145.—PERCOPSIS Agassiz. (152)

532. **Percopsis guttatus** Agassiz. Vn. (519)

---

[1] This subgenus is called *Fario* in the Synopsis, but the type of *Fario* is probably a genuine *Salmo*.

[2] *Salmo gairdneri* is probably the adult sea-run form of *Salmo irideus*.

[3] *Salvelinus rossi* may be omitted from the lists, as no diagnostic characters of importance occur in the description. It may be treated as a very doubtful synonym of *S. oquassa*. *S. naresi* agrees very closely with *S. oquassa*.

[4] *Salvelinus nitidus* may be omitted, as probably identical with *S. stagnalis*. For a description of this species see *Drexel*, Proc. U. S. Nat. Mus., 1884, 255.

Family LI.—STERNOPTYCHIDÆ.[1] (43)

146.—ARGYROPELECUS [2] Cocco.

533. **Argyropelecus hemigymnus** Cocco. O. En.
534. **Argyropelecus olfersi** Cuvier. O. En.

147.—STERNOPTYX [3] Hermann.

535. **Sternoptyx diaphana** Hermann. O. En.

---

[1] A suborder *Iniomi*, to include the *Sternoptychidæ* and *Chauliodontidæ*, has been proposed by Dr. Gill, Proc. U. S. Nat. Mus., 1884, 350. The chief respect in which these families differ from the other *Isospondyli* is in the mode of articulation of the scapular arches, which connect with and impinge on the occiput behind and are otherwise free from the cranium. (Ἰνίον, nape; ὠμός, shoulder.)

Dr. Günther and others have stated that the *Sternoptychidæ* possess a "rudimentary spinous dorsal fin." This appearance is due to the projection of one or more of the neural spines beyond the muscles, and is in no proper sense a rudiment of a fin. (See Gill, l. c., 350.)

[2] ARGYROPELECUS Cocco.

(*Pleurothyris* Lowe.)

(Cocco, Giorn. Sci. Sicil., 1829, fasc. 77, p. 146; type, *Argyropelecus hemigymnus* Cocco.)

Body much elevated and compressed, passing abruptly into the slender tail; no scales, the skin covered with silvery pigment; series of phosphorescent spots along the lower side of the head, body, and tail. Head large, compressed, and elevated, the bones thin but ossified. Cleft of mouth wide, vertical, the lower jaw prominent. Margin of upper jaw formed by the maxillary and premaxillary, both of which have a sharp edge, which is beset with minute teeth; lower jaw and palatine bones with a series of small curved teeth. Eyes large, very close together, lateral, but directed upwards. Angle of preopercle with a spine usually directed downwards. Pectorals well developed; ventrals very small. Humeral arch and pubic bones prolonged into flat pointed processes, which project in the median line of the belly; a series of imbricated scales from the humeral bone to the pubic spine, forming a ventral serratura. Dorsal fin short, median, preceded by a serrated osseous ridge, consisting of several neural spines prolonged beyond the muscles. Adipose fin rudimentary; anal fin short; caudal forked. Gill opening very short, the outer branchial arch extending forward to behind the symphysis of the lower jaw, and beset with very long gill rakers; branchiostegals nine; pseudobranchiæ and air-bladder present. Four pyloric cœca. Small pelagic fishes. (Ἄργυρος, silvery; πέλεκυς, hatchet.)

Argyropelecus hemigymnus Cocco. Depth of body equal to distance between gill-openings and base of caudal; posterior corner of mandible and angle of preopercle each with a small triangular spine; tail without spines; pectoral fin nearly reaching anal. B. 9, D. 7 or 8, A. 11, P. 9, V. 5, L. 2 inches, (*Günther*). Atlantic and Mediterranean in deep water; not rare in the Gulf Stream off Southern New England.

(Cocco, l. c., Cuv. & Val. XXII, 398; Günther, V, 385; Goode & Bean, Bull. Mus. Comp. Zoöl., 1882, 220.)

Argyropelecus olfersi (Cuvier) C. & V. Depth nearly or quite equal to distance from shoulder to root of caudal; tail as deep at base as long. Mandible with a short flat spine at its posterior corner; preopercular spine directed downwards; tail without spines; pectoral fin reaching ventrals. B. 9, D. 9, A. 11, P. 10, V. 6 (*Günther*). Coast of Norway, lately taken in the Gulf Stream, off Southern New England.

(*Sternoptyx olfersi* Cuvier, Règne Animal., ed. 2d, II, 316; Cuv. & Val. XXII, 408; Günther, V, 386; *Pleurothyris olfersi* Lowe, Fish. Madeira, 64.)

[3] STERNOPTYX Hermann.

(Hermann, Naturforscher, 1771, XVI, 8; type *Sternoptyx diaphana* Hermann.)

Trunk much elevated and compressed, the slender tail very short; abdominal out-

Family LII.—CHAULIODONTIDÆ. (44)

148.—CHAULIODUS Bloch & Schneider. (133)
536. Chauliodus sloani Bloch & Schneider. B. Ev. (469)

149.—CYCLOTHONE[1] Goode & Bean.
537. Cyclothone lusca Goode & Bean. B.

150.—SIGMOPS[2] Gill.
538. Sigmops stigmaticus Gill. B.

---

line nearly continuous, in a sigmoid curve; teeth of the jaws in several series, the largest teeth in the inner row; a single spike-like neural spine before dorsal; branchiostegals, 5. Otherwise essentially as in *Argyropelecus*. ($\Sigma\tau\epsilon\rho\nu\sigma\nu$, breast; $\pi\tau\nu\xi$, fold or plait.)

Sternoptyx diaphana Hermann.

Depth equal to distance between tip of snout and base of the very short tail. Interorbital space slightly concave; posterior limb of preopercle bordering hind part of orbit, and descending very obliquely, ending in two points. Pectoral scarcely reaching ventrals, which are very small. B. 5, D. 9, A. 13, P. 10, V. 3. (*Günther.*) Atlantic; lately taken in the Gulf Stream, about lat. 33°.

(Hermann, l. c.; Günther, V, 387; Goode & Bean, Bull. Mus. Comp. Zoöl., 1882, 220.)

[1] CYCLOTHONE Goode & Bean.

(Goode & Bean, Bull. Mus. Comp. Zoöl., 1882, 221; type *Cyclothone lusca* G. & B.)

Body elongate, somewhat compressed (apparently covered with rather large, thin, very caducous scales); lower parts with a series of luminous spots. Head conical; cleft of mouth very wide, oblique extending behind eye, the lower jaw strongly projecting. Maxillary long and slender, sickle-shaped, closely connected with the short premaxillary. Upper jaw with a single series of rather large close-set sharp teeth, about every fourth one slightly longer than the rest, and directed slightly outward. Lower jaw with similar teeth, subequal, directed forward, with a few canines in front. A small patch of minute teeth on vomer; palatines smooth. Eye small, inconspicuous. Gill openings very wide, the membranes free from the isthmus. Gill rakers numerous, long and slender. Pseudobranchiæ none. Branchiostegals (apparently 7 to 9). No air-bladder. Dorsal and anal well developed, opposite each other. No adipose fin. Caudal forked, its peduncle long and slender. Deep-sea fishes of small size, closely related to the European genus *Gonostoma*. ($K\upsilon\kappa\lambda\sigma\varsigma$, round; $\vartheta\omega\nu\eta$, veil.)

Cyclothone lusca Goode & Bean.

Uniform black, the mucous pores inconspicuous. Maxillary extending backward to a distance from tip of snout equal to length of head without snout; eye as long as snout, 7 in head. Distance from snout to dorsal three times length of lower jaw, its base as long as head. Second ray longest, ⅔ base of fin. Insertion of anal under second ray of dorsal, its longest rays a little higher than those of dorsal. Pectoral, 7⅜ in length of body. Distance from snout to ventral twice head; ventral 7 in body. Head, 4⅞; depth, 7¼. D. 1, 11, A. 1, 16, P. 10, V. 5. Gulf Stream, in deep water off south coast of New England, not rare.

(Goode & Bean, Bull. Mus. Comp. Zoöl. 1882, 221.)

[2] SIGMOPS Gill.

(Gill, Proc. U. S. Nat. Mus., 1883, 256; type *Sigmops stigmaticus* Gill.)

No scales or pseudobranchiæ; body elongate, claviform; dorsal short; anal long, the insertions of the two fins opposite each other; teeth moderately elongate, alter-

Order P.—HAPLOMI. (N)

Family LIII.—AMBLYOPSIDÆ. (48)

151.—AMBLYOPSIS De Kay. (153)

539. Amblyopsis spelæus De Kay. Vw. (520)

152.—TYPHLICHTHYS Girard. (154)

540. Typhlichthys subterraneus Girard. Vw. (521)

153.—CHOLOGASTER Agassiz. (155)

541. Chologaster cornutus Agassiz. Vso. (522)
542. Chologaster agassizii Putnam. Vw. (523)
543. Chologaster papillifer Forbes. Vw. (523b.)

Family LIV.—CYPRINODONTIDÆ. (49)

154.—JORDANELLA Goode & Bean. (156)

544. Jordanella floridæ Goode & Bean. Vw. (524)

155.—CYPRINODON Lacépède. (157)

545. Cyprinodon variegatus Lacépède. N. S. (525)
545b. *Cyprinodon variegatus gibbosus* Girard. S. (526)
546. Cyprinodon riverendi[1] Poey. W.
547. Cyprinodon bovinus[2] Girard. Vsw. (526)
548. Cyprinodon eximius[2] Girard. Vsw. (526b.)
549. Cyprinodon latifasciatus Garman. Vsw. (527)
550. Cyprinodon elegans Baird & Girard. Vsw. (528)
551. Cyprinodon californiensis Girard. C? (529)
552. Cyprinodon macularius Girard. R. (530)
553. Cyprinodon mydrus[3] Goode & Bean. S. W.
554. Cyprinodon carpio Günther. (531)

---

nating with short ones, in a row on the maxillaries as well as premaxillaries and mandible. Deep-sea fishes. (Σίγμα, S; οψ, eye.)

Sigmops stigmaticus Gill.

"Its distinct inferior pearly spots, arranged in two rows on each side of the abdomen, are well marked, and the upper have wax-like guttiform spots connected with them below; there is also a broad longitudinal silvery band or sheen." Gulf Stream, lat. 38, at 2,361 fathoms.

(Gill, Proc. U. S. Nat. Mus., 1882, 256.)

[1] *Cyprinodon riverendi* Poey; *Trifarcius riverendi* Poey, Memorias Cuba, II, 306, 1860; *Cyprinodon riverendi* Jordan, Proc. U. S. Nat. Mus., 1884, 109; Key West to Cuba. Very closely related to *C. gibbosus*, but with larger scales (24-12), smaller head and the anal edged with black. The genus *Trifarcius* Poey, of which this species is the type, is founded on the erroneous statement of Valenciennes that *Cyprinodon variegatus* has but five branchiostegals.

[2] A doubtful species, unknown to me.

[3] *Cyprinodon mydrus* Goode & Bean, Proc. U. S. Nat. Mus., 1882, 433; Jordan and Gilbert, Proc. U. S. Nat. Mus., 1884, 110; Pensacola to Key West. A strongly marked and handsome species, possibly identical with *C. carpio*.

156.—CHARACODON[1] Günther.

555. Characodon furcidens Jordan & Gilbert. P.

157.—ADINIA Girard.

556. Adinia multifasciata[2] Girard. S. (545b.)

158.—FUNDULUS Lacépède. (158)

§ *Hydrargyra.*

557. Fundulus majalis[3] Walbaum. N. (532)
558. Fundulus similis Baird & Girard. S. (534)
559. Fundulus parvipinnis Girard. C. P. (536)

§ *Fundulus.*

560. Fundulus zebrinus[4] Jordan & Gilbert. Vsw. (535)

---

[1] CHARACODON Günther.

(Günther, Cat. Fish. Brit. Mus., VI, 1866, 308; type *Characodon lateralis* Günther.)
This genus differs from *Cyprinodon*, chiefly in the presence of a small band of villiform teeth behind the incisors. The incisors are bicuspid or Y-shaped, and the vertical fins are longer than in *Cyprinodon*; fresh waters of Mexico and Central America; two species known. (Χάραξ, a sharp stake; ὀδών, tooth.) *Characodon furcidens* Jordan & Gilbert, Proc. U. S. Nat Mus., 1882, 354; streams tributary to the Gulf of California, and southward; abundant.

[2] The group *Adinia*, defined on page 891 in the Synopsis, may be recognized as a distinct genus, intermediate between *Cyprinodon* and *Fundulus*, having the form of body and restricted gill openings of the former and the dentition of the latter. The single species (*Fundulus xenicus* Jor. & Gilb.) may stand as *Adinia multifasciata*.

[3] *Fundulus swampina*, a doubtful species probably based on a confusion of several species, is here omitted.

[4] *Fundulus zebrinus* is thus redescribed by Professor Gilbert (Bull. Washburn Lab. Nat. Hist., 1, 1884, 15), from specimens taken at Ellis, Kans.:

"Head and body shaped much as in *Fundulus similis*, but the snout somewhat less elongate. Width of preorbital about 6¼ in length of head; eye moderate, 4 to 4¼ in head, 1¾ in interorbital width; posterior margin of orbit in middle of length of head; teeth in both jaws in a villiform band, with the external series much enlarged; interorbital width 2¾ in head; snout 3¼.

"Branchiostegals 5.

"Dorsal fin long and rather low, the base longer and the rays higher in males than in females; origin of dorsal nearly equidistant between snout and margin of caudal, slightly nearer the snout in males, and nearer end of caudal in females; base of dorsal in males 6 to 6¼ in total length, the highest dorsal ray about half head; in females the base is 7¼ in total length. Origin of anal opposite that of dorsal in males, behind it in females; in the latter the anal is sharply angulated, the anterior rays more than thrice the height of the posterior, and more than two-thirds length of head. In males the margins of both dorsal and anal fins are evenly rounded, the anal is the highest, its rays beset with minute white prickles. Oviduct forming a low sheath along base of anterior half of anal. Pectorals not reaching base of ventrals, equaling distance from snout to preopercular margin. Ventrals about reaching vent. Caudal truncate, 1¼ in head.

"Scales very small, in about 60 oblique series from opercle to base of caudal; about 21 in an oblique series from vent upwards to middle of back; no enlarged humeral scale. In males the margins of scales are rough with minute tubercles.

"Head 3¼ to 3¾ in length; depth 4½ to 4¾. D. 14 or 15; A. 13 or 14. L. 3 inches.

"Color: Greenish above, sides and below silvery-white, the sides tinged with sul-

561. **Fundulus seminolis**[1] Girard. Vsw. (537)
562. **Fundulus extensus**,[2] Jordan & Gilbert. P.
563. **Fundulus diaphanus**[3] Le Sueur. Vn. N. (538, 540)
564. **Fundulus confluentus** Goode & Bean. S. (539)
565. **Fundulus adinia** Jordan & Gilbert. Vsw. (541)
566. **Fundulus heteroclitus**[4] Linnæus. N. S. (543)
566 b. *Fundulus heteroclitus grandis* Baird & Girard. S. (543 b.)
567. **Fundulus ocellaris** Jordan & Gilbert. S. (542 b.)
568. **Fundulus vinctus**[5] Jordan & Gilbert. P.

§ *Xenisma* Jordan.

569. **Fundulus catenatus** Storer. Vs. (544)
570. **Fundulus stellifer** Jordan. Vs. (545)

159.—ZYGONECTES Agassiz. (159)

571. **Zygonectes rubrifrons** Jordan. Vse. (546)
572. **Zygonectes henshalli** Jordan. Vsc. (547)
573. **Zygonectes floripinnis** Cope. R. (548)
574. **Zygonectes lineatus** Garman. R. (549).
575. **Zygonectes sciadicus** Cope. Vnw. (555)
576. **Zygonectes notatus** Rafinesque. Vw. (550)
577. **Zygonectes dispar** Agassiz. Vw. (553)
578. **Zygonectes craticula** Goode & Bean. Vse. (553 b.)
579. **Zygonectes zonifer**[6] Jordan & Meek. Vse.
580. **Zygonectes chrysotus**[7] Günther. Vse. (556, 557)
581. **Zygonectes luciæ**[8] Baird. Ve. (558)

160.—LUCANIA Girard. (160)

582. **Lucania venusta** Girard. S. (559)
583. **Lucania parva** Baird & Girard. N. S. (560)
584. **Lucania goodei** Jordan. S. (561)

---

phnr-yellow; the greater part of each scale on back rendered dusky by black points; sides with from 14 to 18 dusky bars from back to ventral region, occasionally meeting on ventral line; these bars are very variable in width, seemingly narrower in females, in which half-bars are frequently inserted between the others; the interspaces are as wide as the bars, or usually wider. Fins yellowish, without distinct markings, in the males all very dusky except the anal."

[1] This species is redescribed by Jordan (Proc. U. S. Nat. Mus , 1884, 322).

[2] *Fundulus extensus* Jordan & Gilbert, Proc. U. S. Nat. Mus., 1882, 355. Cape San Lucas.

[3] *Fundulus menona* appears to be identical with *F. diaphanus*.

[4] *Fundulus nigrofasciatus* seems to be the young of *Fundulus heteroclitus*.

[5] *Fundulus vinctus* Jordan & Gilbert, Proc. U. S. Nat. Mus., 1882, 355. Cape San Lucas.

[6] *Zygonectes zonifer* Jordan & Meek, Proc. U. S. Nat. Mus., 1884. Allamaha R., Ga.

[7] ? *Fundulus cingulatus* Cuv. & Val. = *Haplochilus chrysotus* Günther = *Fundulus zonatus* C. & V., not *Esox zonatus* Mitchill, which is a young *Fundulus*. For descriptions of this species see Jordan & Gilbert, Proc. U. S. Nat. Mus., 1882, 586, and Jordan, op. cit., 1884, 320. It is best to use the name of *chrysotus* for this species, as *cingulatus* cannot be positively identified, and *zonatus* was originally given to some other fish.

[8] The description of *Zygonectes cingulatus* given in the Synopsis (p. 342) belongs to this species. It is probably distinct from *Z. chrysotus*, as the latter has no dorsal ocellus in either sex.

161.—GAMBUSIA Poey. (161)

585. Gambusia patruelis[1] Baird & Girard. Vs. (551, 552, 562)
586. Gambusia humilis[2] Günther. Vsw. (554, 463)
587. Gambusia arlingtonia[3] Goode & Bean. Vse. (564)
588. Gambusia affinis[3] Baird & Girard. Vsw. (565)
589. Gambusia nobilis[3] Baird & Girard. Vsw. (566)
590. Gambusia senilis[3] Girard. Vsw. (566 b.)

162.—MOLLIENESIA Le Sueur. (162)

591. Mollienesia latipinna[4] Le Sueur. S. (567, 567 b.)

163.—PŒCILIA Bloch & Schneider. (163)

592. Pœcilia couchiana Girard. Vsw. (568)

164.—HETERANDRIA[5] Agassiz. (164)

593. Heterandria formosa Agassiz. Vse. (164)
594. Heterandria occidentalis Baird & Girard. R. (570)
595. Heterandria ommata[6] Jordan. Vse.

Family LV.—UMBRIDÆ. (50)

165.—UMBRA Müller. (169)

596. Umbra limi Kirtland. Vnw. (571)
596 b. *Umbra limi pygmæa* DeKay. Ve.

Family LVI.—ESOCIDÆ. (51)

166.—ESOX Linnæus. (167)

§ *Picorellus* Rafinesque.

597. Esox americanus Gmelin. Ve. (573)
598. Esox vermiculatus Le Sueur. Vw. (574)
599. Esox reticulatus[7] Le Sueur. Ve. (575)

---

[1] *Zygonectes atrilatus, Zygonectes inurus, Haplochilus melanops, Gambusia holbrooki*, and probably *Gambusia arlingtonia* also, are identical with *Gambusia patruelis*.

[2] *Gambusia humilis* Günther=*Zygonectes brachypterus* Cope, seems to be distinct from *Gambusia patruelis*. It abounds in the streams of Texas, and may be known at once from *G. patruelis* by the absence of the black suborbital spot.

[3] Doubtful species, unknown to me.

[4] *Mollienesia lineolata* is identical with *M. latipinna*.

[5] The name *Heterandria* Agassiz, Amer. Journ. Sci. Arts., 1853, as now restricted is identical with *Girardinus*, and must supersede this later name. The type is *Heterandria formosa* Agassiz. As originally defined, both *Gambusia* and *Girardinus* were included in *Heterandria*. See Jordan & Meek, Proc. U. S. Nat. Mus., 1884, 236.

[6] *Heterandria ommata* Jordan, Proc. U. S. Nat. Mus., 1884, 323. Indian R., Florida.

[7] This species should stand as *Esox vermiculatus*, instead of *Esox salmoneus* or *Esox umbrosus*.
To the synonymy add:
(*Esox vermiculatus, Esox lineatus*, and ? *Esox lugubrosus* Le Sueur MSS. in Cuv. & Val., XVIII, 333, 335, 338, 1846.)

§ *Esox.*
600. Esox lucius Linnæus. Eu. Vn. (576)
§ *Mascalongus* Jordan.
601. Esox nobilior Thompson. Vn. (577)

Order Q.—XENOMI.[1]

Family LVII.—DALLIIDÆ.

167.—DALLIA Bean. (166)
602. Dallia pectoralis Bean. Y. (572)

Order R.—COLOCEPHALI.[2]

Family LVIII.—MURÆNIDÆ. (52.)

168.—MURÆNOBLENNA[3] Lacépède.
603. Muraenoblenna nectura Jordan & Gilbert. P.

169.—MURÆNA Linnæus. (168)
604. Muraena retifera Goode & Bean. S. (578)
605. Muraena pinta[4] Jordan & Gilbert. P.

170.—SIDERA Kaup.
606. Sidera castanea[5] Jordan & Gilbert. P.
607. Sidera mordax Ayres. C. (579)
608. Sidera dovii[6] Günther. P.
609. Sidera ocellata Agassiz. S. (580)

---

[1] The genus *Dallia*, although agreeing in many external characters with *Umbra*, has very little affinity with that group or any other of our fishes. Its skeleton is so peculiar in structure that it has been taken by Dr. Gill as the representative of a peculiar order or suborder, *Xenomi*, which is thus defined:

"Teleosts with the scapular arch free from the cranium laterally and only abutting on it behind, coracoids represented by a simple cartilaginous plate without developed actinosts, and with the intermaxillary and supramaxillary bones coalescent."
($\Xi \acute{\epsilon} \nu o \varsigma$, strange; $\tilde{\omega} \mu o \varsigma$, shoulder.)

[2] Order *Colocephali* Cope, Trans. Am. Philos. Soc., 1871, 456 (includes the *Murænidæ*).

[3] MURÆNOBLENNA Lacépède.
(*Gymnomuræna* Günther, not of Lacépède, as restricted by Kaup.)
(Lacépède, His. Nat. Poiss., V, 652, 1803; type *Muraenoblenna olivacea* Lacépède.)
This genus differs from *Muraena* chiefly in the reduction of the fins to a short fold, surrounding the tail. Posterior nostrils not tubular. Gape, moderate. Tropical seas. ($M \upsilon \rho \alpha \acute{\iota} \nu \alpha$, eel; $\beta \lambda \epsilon \nu \nu \alpha$, slime. "Blenna en grec, signifié mucosité." Lacépède.) *Muraenoblenna nectura = Gymnomuraena nectura* Jordan & Gilbert, Proc. U. S. Nat. Mus., 1882, 356. Cape San Lucas.

[4] *Muraena pinta* Jordan & Gilbert, Proc. U. S. Nat. Mus., 1881, 345. Gulf of California and southward.

[5] *Sidera castanea* Jordan & Gilbert, Proc. U. S. Nat. Mus., 1883, 208. Mazatlan and southward. In this paper is an analysis of the characters of the species of *Sidera* found on the Pacific coast of America.

[6] *Muraena dovii* Günther, VIII, 103, 1870; = *Muraena pintita* Jordan & Gilbert, Proc. U. S. Nat. Mus., 1881, 346; 1883, 209. Mazatlan to Gallapagos Islands.

610. **Sidera funebris**[1] Ranzani. P. (580 b.)
611. **Sidera moringa** Cuvier. P. (580 c.)

Order S.—ENCHELYCEPHALI.[2] (O.)

Family LIX.—CONGRIDÆ.[3] (53 part.)

171.—ICHTHYAPUS[4] Barneville.
612. **Ichthyapus selachops** Jordan & Gilbert. P.

172.—LETHARCHUS Goode & Bean. (168 b.)
613. **Letharchus velifer** Goode & Bean. S. (580 b.)

173.—CALLECHELYS[5] Kaup. (169)
614. **Callechelys scuticaris** Goode & Bean. S. (581)
615. **Callechelys teres** Goode & Bean. S. (581 b.)
616. **Callechelys bascanium**[6] Jordan. W.

---

[1] The species called in the Synopsis (p. 895) *Murœna afra* should stand as *Murœna* or *Sidera funebris*.

In life this species is bright yellowish green, with some oblique dark streaks on the fins. It reaches a very large size and is much dreaded by fishermen. To its synonymy add: *Gymnothorax funebris* Ranzani, Nov. Comm. Ac. Sci. Inst. Bonon., IV, 1840, 76; *Murœna lineopinnis* Richardson, Voy. Erebus & Terror, 1844, 89; *Murœna infernalis* Poey, Memorias Cuba, II, 347, 1861; *Murœna afra* Günther, IX, 123; apparently not *Gymnothorax afer*, Bloch, Ausl. Fische, 1797, IX, 85, tab. 417, a fish from Guinea, described as being brown, marbled, and banded with white. The present species is always unicolor, green in life, and brown in spirits.)

[2] *Enchelycephali* Cope, Trans. Am. Philos. Soc., 1871, 455.

[3] The family of *Anguillidæ*, as given in the text, is not a natural one. For the present we may subtract the aberrant genera *Anguilla* and *Simenchelys*, leaving the remaining genera in one group, *Congridæ*.

[4] ICHTHYAPUS Barneville.

(*Ophisuraphis* Kaup; *Apterichthys* Duméril.)

(Barneville, Revue Zoologique, 1847, 219; type *Ichthyapus acutirostris* Barneville.)

This genus differs from *Ophichthys* chiefly in the entire absence of fins. The snout projects beyond the small mouth, giving a shark-like physiognomy, and the teeth are small, mostly uniserial. ("Ἰχθύς, fish; ἄπους, without feet.) *Ichthyapus selachops* = *Apterichthys selachops* Jordan & Gilbert, Proc. U. S. Nat. Mus., 1882, 356. Cape San Lucas.

[5] *Callechelys* Kaup (see Synopsis, p. 897), is distinguished from *Cœcula* by the development of the dorsal fin, which begins on the head. In *Cœcula* (*Sphagebranchus*), it begins behind the gill opening.

[6] *Callechelys bascanium* Jordan.

Dark brown, nearly uniform; fins a little paler. Body extremely slender, subterete, its greatest depth little more than two-fifths length of head; head short; snout 7 in head; mouth very small, the lower jaw thin, included, not extending to the anterior nostril, which is in a short tube; teeth short, subconic, bluntish, a little unequal, their points directed backwards; lower teeth nearly uniserial; upper teeth uniserial laterally, partly biserial anteriorly; vomerine teeth forming a rhombic patch. Eye moderate, its length more than half that of snout, its center nearly over middle of upper jaw; cleft of mouth 3¾ in length of head. Gill openings vertical, about as wide as isthmus; its upper edge on level of upper base of pectoral; pectoral developed, small, a little broader than long, nearly as long as snout; dorsal fin very low, beginning at a point midway between front of eye and gill opening; anal similar to dorsal.

174.—OPHISURUS[1] Lacépède. (170 b.)

617. Ophisurus acuminatus[2] Gronow. W. (584 b.)
618. Ophisurus xysturus[3] Jordan & Gilbert. P.

175.—OPHICHTHYS[1] Ahl. (170)

619. Ophichthys miurus[4] Jordan & Gilbert. P.
620. Ophichthys triserialis Kaup. C. P. (583)
621. Ophichthys ocellatus Le Sueur. P. (584)
622. Ophichthys guttifer[5] Bean & Dresel. W.
623. Ophichthys macrurus Poey. W. (583 b.)
624. Ophichthys chrysops Poey. W. (583 c)
625. Ophichthys zophochir[6] Jordan & Gilbert. P.
626. Ophichthys schneideri[7] Steindachner. W. (582)
627. Ophichthys intertinctus[8] Richardson. W.

Head 11¼ in distance from top of snout to vent; head and trunk a little longer than tail. Length of type, 31 inches; head, 1⅞; trunk, 14⅘. Egmont Key, Florida; distinguished from *C. teres* by the very short head.

(*Cæcula buscanium* Jordan, Proc. Ac. Nat. Sci., Phila., 1884, 43.)

[1] For a discussion of the correct application of the names *Ophichthys*, *Ophisurus*, and *Cæcula* see Jordan & Gilbert, Proc. U. S. Nat. Mus., 1884, 648.

[2] As stated in the Synopsis, p. 974, the name *acuminatus* should supersede *longus* for this species.

[3] *Ophichthys xysturus* Jordan & Gilbert, Proc. U. S. Nat. Mus., 1881, 346. Mazatlan to Panama.

[4] *Ophichthys miurus* Jordan & Gilbert, Proc. U. S. Nat. Mus., 1882, 357. Cape San Lucas.

[5] *Ophichthys guttifer* Bean & Dresel.

Allied to *O. ocellatus* Le Sueur. Greatest depth equal to distance from angle of mouth to tip of snout. Dorsal fin beginning at a distance behind vertical from tip of pectoral equal to length of snout. Pectoral nearly 3½ in head; head 8 in total length, 2⅞ in trunk. Eye 1½ in snout; 9 in head. Twenty-one or 22 small white spots along median line. Gulf of Mexico. (Bean & Dresel, Proc. Biol. Soc., Washington, II, 1884, 99.)

[6] *Ophichthys zophochir* Jordan & Gilbert, Proc. U. S. Nat. Mus., 1881, 347. Mazatlan.

[7] The specimens which we have referred to *Ophichthys punctifer* (*mordax*) belong rather to *Ophichthys schneideri* Steindachner.

Yellowish brown; head with small dark brown elongate spots; sides with about three rows of rather large oval spots, the lower disappearing behind the vent, number of rows becoming greater anteriorly; broad half spots along upper margin of dorsal, and bordered with blackish. Head 3½ in trunk; snout conical, blunt anteriorly. Cleft of mouth very long, 2 in head; eye 11; snout 7. Teeth in both jaws in two rows, those of the outer row in both very sharp, unequal, some of them quite long, those of the inner row smaller and subequal; *vomerine teeth* rather small, *in two rows*, diverging forward; one or two long canines in front, behind the two series of the upper jaw. Both nostrils with short tubes. Pectoral 4 in head; dorsal beginning about 1½ eye's diameters behind the point of the pectoral. Tail longer than the rest of the body by 1½ head's lengths. (*Steindachner*.) West Indies, occasionally taken from the stomachs of Red Snappers at Pensacola. Apparently distinct from *O. punctifer* (=*O. mordax*), having the vomerine teeth in two rows instead of three.

Crotalopsis mordax Goode & Bean, Proc. U. S. Nat. Mus., 1879, 154; not *Macrodonophis mordax* Poey; Steindachner, Ichth. Beitr., VIII, 67, 1879; Jordan & Gilbert, Proc. U. S. Nat. Mus., 1883, 143.)

[8] *Ophichthys intertinctus*.

Dark brown above, paler below; sides and back with about three rows of large ovate brown spots, somewhat irregular in size and position, those of the upper row smallest, the large and small ones of the lower rows somewhat alternating. Spots on head small and numerous. Dorsal with an interrupted dark margin; anal with

176.—MYRICHTHYS Girard. (171)

628. **Myrichthys tigrinus** Girard. C. (585)

177.—MYROPHIS Lütken. (171 b.)

629. **Myrophis lumbricus** Jordan & Gilbert. S. (585 b.)
630. **Myrophis punctatus**[1] Lütken. W. (585 c.)
631. **Myrophis vafer**[2] Jordan & Gilbert. P.
632. **Myrophis egmontis**[3] Jordan. W.

178.—NEOCONGER Girard. (172)

633. **Neoconger mucronatus** Girard. W. (586)

179.—NETTASTOMA[4] Rafinesque.

634. **Nettastoma procerum** Goode & Bean. B.

---

a darker edge; pectorals blackish. Gill openings wide, the isthmus rather narrow; head 3½ in trunk. Cleft of mouth very wide, nearly half length of head. Teeth sharply pointed, with a few large fixed canines in both jaws, and one or two larger ones in front of upper jaw; about 4 moderate canines near front of lower jaw; teeth in both jaws in double series, those of the inner series in the upper jaw depressible. Vomer with a double series confluent behind. Eye small, 1½ in snout, which is about 6½ in head. Pectoral about 5 in head. Dorsal commencing a little behind end of pectoral. Tail rather longer than rest of body. West Indies, north to Egmont Key, Florida.

(*Ophisurus intertinctus* Richardson, Ereb. & Terr. Fish., 102; *Echiopsis intertinctus* Kaup, Apodes, 13, 1858; Günther, VIII, 57; *Ophichthys intertinctus* Jordan, Proc. Ac. Nat. Sci. Phila., 1884, 43.)

[1] *Myrophis punctatus* Lütken=*Myrophis microstigmius* Poey. To the synonymy, add—
(Lütken, Vid. Med. Naturh. Foren. Kjobenh., 1851, 1; *Myrophis longicollis* Kaup, Apodes, 30, 1858; Jordan, Proc. Ac. Nat. Sci. Phila., 1883, 282; not of Günther, VIII, 51,=*M. vafer* Jor. & Gilb.)

[2] *Myrophis vafer* Jordan & Gilbert, Proc. U. S. Nat. Mus., 1882, 645. Guaymas to Panama.

[3] *Myrophis egmontis* Jordan.

Dark brown, apparently uniform, somewhat paler below; head small, slender, moderately pointed; anterior nostril in a short tube; posterior, large, labial directly behind it; cleft of mouth rather short, extending to beyond the rather large eye, which is more than half the length of the snout; cleft of mouth, 3⅙ in head; teeth on both jaws subequal, pointed, slightly compressed, arranged in single series, those of both jaws directed somewhat backward; the lower teeth larger and more oblique than the upper; about four small fixed canines in front of upper jaw; no teeth on vomer in two specimens examined; tongue not free; lower jaw considerably shorter than upper, its edge considerably curved, concave in outline. Nape somewhat elevated; top of head with large pores. Head 5¼ in distance from snout to vent; head and trunk a little shorter than tail; body slender, its greatest depth a little more than length of gape. Pectoral short and broad, slightly longer than snout; the gill opening short, oblique, extending downward and backward from near the middle of the base of the pectoral. Dorsal fin beginning behind vent, at a distance about equal to length of gape; the fin very low in front, becoming gradually higher towards the tip of tail; anal low, but well developed, considerably higher than dorsal, highest anteriorly, uniting with the dorsal around the tail. Length, 15 inches. Egmont Key, Florida.

(Jordan, Proc. Ac. Nat. Sci. Phila., 1884, 44.)

[4] NETTASTOMA Rafinesque.

(*Hyoprorus* Kölliker; larva.)

(Rafinesque, Caratteri di Alcuni Nuovi Generi, &c., 1810, 66; type *Nettastoma melanura* Raf.)

Scaleless. Tail tapering into a point. Snout much produced, depressed; jaws and

180.—MURÆNESOX[1] McClelland.
635. **Muraenesox coniceps** Jordan & Gilbert. P.

181.—CONGER[2] Cuvier. (174)
636. **Conger conger** Linnæus. N. S. W. Eu. P. (588)
637. **Conger caudicula** Bean. W. (588 b.)

Family LX.—ANGUILLIDÆ.
182.—ANGUILLA[3] Thunberg. (173)
638. **Anguilla anguilla rostrata** De Kay. V. N. S. W. (587)

vomer with bands of cardiform teeth, those along the median line of the vomer being somewhat the larger. Vertical fins well developed, the dorsal commencing behind gill opening; no pectorals. Gill openings moderate. Nostrils on upper surface of head, valvular, the anterior near end of snout, the posterior above anterior angle of eye. Air bladder present. (Νηττα, duck; στόμα, mouth.)

Nettastoma procerum Goode & Bean.

Body extremely elongate, compressed, especially so posteriorly, the tail tapering to a very attenuate point. Head slender, conical, the jaws somewhat depressed, the upper heavier and thicker, projecting beyond the lower a distance equal to the diameter of the eye. Numerous pores on both jaws and on the nape. Snout with a slender filamentous tip, twice as long as the eye. Teeth arranged as in *N. melanurum*, but excessively small. Dorsal commencing above gill opening. Insertion of anal at a distance from snout equal to 3¾ times length of head. Tail twice as long as head and body. Lateral line well developed, in a deep furrow. Height of dorsal and anal about half depth of body, brownish; peritoneum black. (Gulf Stream, in deep water, at about lat. 34°. (*Goode & Bean*.)

(Goode & Bean, Bull. Mus. Comp. Zoöl., 1883, 224.)

[1] MURÆNESOX McClelland.

(*Cynoponticus* Costa.)

Form of *Conger*: Body scaleless; snout long; posterior nostrils opposite upper part of eye; tongue not free; jaws with several series of small, close-set teeth, with canines in front; vomer with several series of strong teeth, those of the median series enlarged and usually compressed; gill openings wide; pectorals well developed; dorsal beginning above the gill opening, continuous with the anal around the tail. Large eels of the tropical seas.

Muraenesox coniceps Jordan & Gilbert, Proc. U. S. Nat. Mus., 1881, 348. Mazatlan to Panama.

[2] The name *Conger* should probably be retained for this genus. It does not appear to be entirely certain that *Leptocephalus morrisi* is a larval Conger. *Echelus* Rafinesque (1810) is based in part on Congers, but most of the numerous typical species remain unidentified.

[3] Mr. S. E. Meek (Bull. U. S. Fish Comm., 1883, 430), after a careful comparison of American and European eels, concludes that "in American specimens the dorsal fin is proportionately farther from the end of snout, making the distance between front of dorsal and front of anal a little shorter than in European specimens. Otherwise no permanent difference seems to exist. We should not, therefore, in my opinion, consider the two as distinct species, but rather as geographical varieties of the same species."

In *A. rostrata*, according to Mr. Meek, the distance from tip of snout to front of dorsal is, on an average, .33¼ of the length; the distance from front of dorsal to front of anal, .09¾, or less than length of head (.12¼).

In the European *Anguilla anguilla* the first distance is .30¼, the second, .13¾, or a little more than length of head (.13¼). Cuban specimens (*Anguilla cubana* Kaup) agree fully with *A. rostrata*, as also Texan ones (*Anguilla "tyrannus"* or "*texana*").

Probably our eel should be regarded as a subspecies (*rostrata*) of *A. anguilla*.

Family LXI.—SIMENCHELYIDÆ.

183.—SIMENCHELYS Gill. (174)

639. Simenchelys parasiticus Gill. B. (589)

Family LXII.—SYNAPHOBRANCHIDÆ. (54)

184.—SYNAPHOBRANCHUS Johnson. (176)

640. Synaphobranchus pinnatus Gronow. B. (590)

185.—HISTIOBRANCHUS[1] Gill.

641. Histiobranchus infernalis Gill. B.

Family LXIII.—NEMICHTHYIDÆ Richardson. (56)

186.—NEMICHTHYS Richardson. (178)

642. Nemichthys scolopaceus Richardson. B. (592)
643. Nemichthys avocetta Jordan & Gilbert. B. C. (593)

187.—LABICHTHYS[2] Gill & Ryder.

644. Labichthys carinatus[3] Gill & Ryder. B.
645. Labichthys elongatus[4] Gill & Ryder. B.

---

[1] HISTIOBRANCHUS Gill.

(Gill, Proc. U. S. Nat. Mus., 1883, 255; type, *Histiobranchus infernalis* Gill).

"Synaphobranchid, with the dorsal fin protracted almost as far forward as the base of the pectoral fin, and an isolated small patch of teeth on the vomer, behind that on its head." ($Ἱστίον$, sail, i. e., dorsal fin; $βραγχος$, gill; dorsal commencing above gill opening).

Histiobranchus infernalis Gill, Proc. U. S. Nat. Mus., 1882, 255. Gulf Stream, latitude 35°, at a depth of 1,731 fathoms.

[2] LABICHTHYS Gill & Ryder.

(Gill & Ryder, Proc. U. S. Nat. Mus., 1883, 261; type, *Labichthys carinatus* Gill & Ryder.)

"*Nemichthyids* with the head behind the eyes, contracted, with very attenuated jaws, the branchiostegous membrane connected to the throat, and the branchial apertures limited to the sides, with small conical teeth in a band along the vomer, and otherwise dentition of *Nemichthys*, a black epidermis, and the tail abruptly truncated. ($Λαβίς$, a pair of forceps; $ἰχθύς$, fish.) This genus and the two which follow are very insufficiently described. In none of them is the character of the posterior dorsal rays described.

[3] *Labichthys carinatus* Gill & Ryder, Proc. U. S. Nat. Mus., 1883, 261. Gulf Stream, latitude 41°, at 906 fathoms.

[4] *Labichthys elongatus* Gill & Ryder, l. c., 1883, 262. Gulf Stream, latitude 39°, at 1,628 fathoms.

188.—SPINIVOMER[1] Gill & Ryder.

646. Spinivomer goodei Gill & Ryder. B.

189.—SERRIVOMER[2] Gill & Ryder.

647. Serrivomer beani Gill & Ryder. B.

Order T—LYOMERI.[3]

Family LXIV.—SACCOPHARYNGIDÆ. (55)

190.—SACCOPHARYNX Mitchill. (177)

648. Saccopharynx ampullaceus[4] Harwood. B. (591)

Family LXV.—EURYPHARYNGIDÆ.[5]

---

[1] SPINIVOMER Gill & Ryder.

(Gill & Ryder, Proc. U. S. Nat. Mus., 1883, 261; type, *Spinivomer goodei* G. & R.)

"*Nemichthyids* with a rectilinear occipitorostral outline, with very attenuated jaws, high mandibular rami, the branchial aperture nearly confluent, enlarged acute conic teeth in a median row on the vomer, and with a silvery epidermis and filiform tail."

(Latin, *spina*, spine; *vomer*, vomer.)

Spinivomer goodei Gill & Ryder, l. c., 261. Gulf Stream, latitude 38°, at 2,361 fathoms.

[2] SERRIVOMER Gill & Ryder.

(Gill & Ryder, Proc. U. S. Nat. Mus., 1883, 260; type, *Serrivomer beani* G. & R.)

"*Nemichthyids* with the head behind eyes of an elongated parallelogramic form, with moderately attenuated jaws, branchiostegal membrane confluent at posterior margin, but with the branchial aperture limited by an isthmus except at the margin, and with lancet-shaped vomerine teeth in a crowded (sometimes doubled) row."

(Latin, *serra*, saw; *vomer*, vomer.)

Serrivomer beani Gill & Ryder, l. c., 261. Gulf Stream, latitude 41°, at 855 fathoms.

[3] Order T.—LYOMERI.

"Fishes with five branchial arches (none modified as branchiostegal or pharyngeal) far behind the skull, an imperfectly ossified cranium articulating with the first vertebra by a basioccipital condyle alone, only two cephalic arches, both freely movable, (1) an anterior dentigerous one, the palatine, and (2) the suspensorial, consisting of the hyomandibular and quadrate bones, without maxillary bones or distinct bony elements to the mandible, with an imperfect scapular arch remote from the skull, and with separately ossified but imperfect vertebræ." (Gill & Ryder.)

Two families are recognized (*Saccopharyngidæ* and *Eurypharyngidæ*), deep-sea fishes of remarkable appearance, allied to the eels. The species are little known, and are possibly all forms of a single one. (λvos, loose; $\mu\epsilon\rho os$, part or segment.) (*Lyomeri* Gill & Ryder, Proc. U. S. Nat. Mus., 1883, 263.)

[4] The name *Saccopharynx flagellum* was not given by Mitchill, but by Cuvier (Règne Animal, Ed. II) in 1829. The name *ampullaceus* of Harwood has therefore priority, it really referring to the same species. For an exhaustive discussion of our knowledge of *Saccopharynx* and its relationships see Gill, Proc. U. S. Nat. Mus., 1884, 48.

[5] The family *Eurypharyngidæ* is thus defined by Gill & Ryder:

"*Lyomeri* with the head flat above and with a transverse rostral margin, at the outer angles of which the eyes are exposed, with the eyes excessively elongated backwards and the upper parallel and closing against each other as far as the articulation

191.—GASTROSTOMUS[1] Gill & Ryder.

649. **Gastrostomus bairdii** Gill & Ryder. B.

Order U.—OPISTHOMI. (P)

Family LXVI.—PTILICHTHYIDÆ.[2] (56 b.)

192.—PTILICHTHYS Bean. (179)

650. **Ptilichthys goodei** Bean. A. (594.)

Family LXVII.—NOTACANTHIDÆ.

193.—NOTACANTHUS Bloch. (180)

651. **Notacanthus chemnitzi** Bloch. G. B. (595)
652. **Notacanthus phasganorus** Goode. B. (595 f.)
653. **Notacanthus analis**[3] Gill. B.

---

of the two suspensorial bones, with minute teeth in both jaws, with a short abdomen and long, attenuated tail, branchial apertures narrow and very far behind, dorsal and anal fins continued nearly to the end of the tail, and minute pectoral fins.

"The mandibular rami are exceedingly narrow and slender, but the jaws are extremely expansible and the skin is correspondingly dilatable, consequently an enormous pouch may be developed. Inasmuch as the slenderness and fragility of the jaws and the absence of raptatorial teeth preclude the idea of the species being true fishes of prey, it is probable that they may derive their food from the water which is received into the pouch by a process of selection of the small or minute organisms therein contained." The skin of the pouch has a peculiar velvety appearance, like the wing membrane of a bat. Two species are known, provisionally referred to two genera, *Eurypharynx pelecanoides* Vaillant and *Gastrostomus bairdii*. Both are from great depths in the sea, the former having been taken by the "Travailleur," in 1882, off the coast of Morocco.

(*Eurypharyngidæ* Gill & Ryder, Proc. U. S. Nat. Mus., 1883, 264.)

[1] GASTROSTOMUS Gill & Ryder.

Gill & Ryder, Proc. U. S. Nat. Mus., 1883, 271; type *Gastrostomus bairdii* G. & R.

This genus is supposed to be distinguished from *Eurypharynx* by the following characters: Cranium short, nearly as broad as long; dentigerous bones almost seven times length of cranium; jaws with minute, acute, conic teeth depressed inwards, in a very narrow band; no enlarged teeth at tip of mandible; tail with a rayless membrane under its tip. (Γαστηρ, stomach; στόμα, mouth.)

(*Gastrostomus bairdii* Gill & Ryder, l. c., 1883, 271. Gulf Stream, lat. 40°, in deep water.)

Eurypharynx pelecanoides (Vaillant, Comptes Rendus Acad. Sci. Paris, 1882, 1239) is supposed to differ in having the "cranium prolonged backwards, the dentigerous bones little more than three times as long as the cranium; faint dentary granulations on both jaws and at the extremity of the mandible two hooked teeth; the tail ending in a point." It is not unlikely that the two species may prove identical.

[2] It is almost certain that *Ptilichthys* has little relation to the *Mastacembelidæ*. It should probably be regarded as a distinct family, *Ptilichthyidæ*, but whether this family belongs to the *Opisthomi* or to the *Acanthopteri* cannot be ascertained without examination of the skeleton.

[3] *Notacanthus analis* Gill. Proc. U. S. Nat. Mus. 1883, 255. Gulf Stream, latitude 40° at a depth of 548 fathoms.

Order V.—SYNENTOGNATHI. (Q)

Family LXVIII.—BELONIDÆ.[1] (57 pt.)

194.—TYLOSURUS[2] Cocco. (181)

654. **Tylosurus hians** Cuv. & Val. W. (696)
655. **Tylosurus fodiator**[3] Jordan & Gilbert. P.
656. **Tylosurus crassus**[4] Poey. W. (600 b.)
657. **Tylosurus caribbæus** Le Sueur. W. (597)
658. **Tylosurus notatus** Poey. W. (598)
659. **Tylosurus sagitta**[5] Jordan & Gilbert. W.
660. **Tylosurus marinus** Bloch & Schneider. N. S. (599)
661. **Tylosurus exilis** Girard. C. (600)
662. **Tylosurus stolzmanni**[6] Steindachner. P.

[1] According to Dr. Gill the structure of the skeleton in *Belone*, *Tylosurus* and *Potamorrhaphis* differs so much from that of the other *Scomberesocidæ* that these genera should be placed in a distinct family, *Belonidæ*.

[2] The identification of our species of *Tylosurus* may be aided by the following key:

a. Body strongly compressed, somewhat band-like, about twice as deep as broad; beak slender, the upper jaw strongly arched at base; dorsal and anal very long, the posterior rays elevated: D. 24; A. 25..........HIANS.
aa. Body subcylindrical, or not greatly compressed.
  b. Dorsal and anal long, each with 20 or more rays, their posterior rays prolonged in the young, short in the adult; scales small; beak strong, with large teeth; lateral line passing into a dark-colored keel on tail, no bluish lateral band; size large.
    c. Beak very strong, not twice as long as rest of head; body comparatively stout; depth about 14.
      d. Dorsal rays about 19; anal 17. ....FODIATOR.
      dd. Dorsal rays about 23. A. 23..................................CRASSUS.
    cc. Beak twice or more length of rest of head; body comparatively slender; depth about 18, D. about 25, A. about 24..............CARIBBÆUS.
  bb. Dorsal and anal short, each with less than 20 rays; the last rays not prolonged; beak long and slender; sides with a bluish lateral band; size small.
    e. Caudal peduncle posteriorly compressed, the lateral line not dark and not forming a keel.
      f. Body very broad, robust; dorsal very short, its lobe orange-red in life; maxillary hidden by preorbital. D. 13; A. 14.............NOTATUS.
      ff. Body very slender, subterete; dorsal moderate, not red; maxillary not hidden by preorbital. Eye small. D. 14, A. 16...........SAGITTA.
    ee. Caudal peduncle posteriorly depressed; lateral line forming a slight keel which is blackish in color; eye rather large: D. 15; A. 18..MARINUS.
    eee. Caudal peduncle depressed, with a strong keel; maxillary not entirely hidden. D. 15 or 16; A. 17.
      g. Pectorals plain olivaceous; dorsal and anal lobe pale.........EXILIS.
      gg. Pectorals abruptly black at tip; dorsal and anal lobes blackish.....
STOLZMANNI.

[3] *Tylosurus fodiator* Jordan & Gilbert, Proc. U. S. Nat. Mus., 1881, 459. Mazatlan.

[4] *Belone crassa* Poey, Memorias Cuba, II, 1860, 291 = *Tylosurus gladius* Bean, Proc. U. S. Nat. Mus., 1882, 430 = *Tylosurus crassus* Jordan, Proc. U. S. Nat. Mus., 1884, 112 (not *Belone jonesi* Goode). Pensacola southward.

[5] *Tylosurus sagitta* Jordan & Gilbert, Proc. U. S. Nat. Mus., 1884, 25. Key West.

[6] *Belone stolzmanni* Steindachner, Ichthyol. Beiträge, VII, 21, 1878 = *Tylosurus sierrita* Jordan & Gilbert, Proc. U. S. Nat. Mus., 1881, 458. Gulf of California to Peru.

195.—SCOMBERESOX Lacépède. (182)

663. Scomberesox saurus Walbaum. N. S. O. Eu. (601)
664. Scomberesox brevirostris Peters. C. (602)

196.—HEMIRHAMPHUS Cuvier. (183)

665. Hemirhamphus unifasciatus[1] Ranzani. W.
666. Hemirhamphus roberti[2] Cuv. & Val. S. P. (603)
667. Hemirhamphus rosæ Jordan & Gilbert. C. (604)
668. Hemirhamphus pleei[3] Cuv. & Val. S. W. P. (604 b.)

197.—EULEPTORHAMPHUS Gill. (183 b.)

669. Euleptorhamphus longirostris Cuvier. O. (605)

198.—CHRIODORUS Goode & Bean. (183 c.)

670. Chriodorus atherinoides Goode & Bean. W. (605 b.)

199.—PAREXOCŒTUS Bleeker.

671. Parexocœtus mesogaster[4] Bloch. W. S. (607 b.)

200.—HALOCYPSELUS Weinland. (184)

672. Halocypselus evolans[5] Linnæus. S. (606; 607)

---

[1] *Hemirhamphus unifasciatus* Ranzani. Clear greenish with bluish luster; a silvery lateral band; no red on fins; tip of lower jaw scarlet. Very close to *H. unifasciatus*, differing chiefly in the shorter beak, and the less compressed and more robust body. Lower jaw from end of upper jaw 6 to 7 in total length from its tip to base of caudal, (4¼ in *H. roberti*) its length always less than that of rest of head; head with lower jaw, 3; body half deeper than broad; premaxillaries broader than long; eye less than interorbital width, ⅔ postorbital part of head; ventrals midway between eye and base of caudal; dorsal and anal densely scaly; back broad. Head 4⅝, depth 6¼. D. 12 to 14, A. 15, lat. l. 52, length 12 inches. Florida Keys to Cuba and Panama, representing *H. roberti* southward.

Hemirhamphus unifasciatus Ranzani, Comm. Inst. Bon., 1842, V. 326, tab. 25; not of most recent authors; ? *Hemirhamphus picarti* Cuv. & Val. XIX, 1-46, 25 (*Hemirhamphus richardi* Cuv. & Val., XIX, 1-46, 26; *Hemirhamphus fasciatus* Poey, Memorias Cuba, II, 299, 1-60, not of Bleeker; *Hemirhamphus poeyi* Günther, VI, 262).

[2] The species called in the text *Hemirhamphus unifasciatus* should stand as *Hemirhamphus roberti* Cuv. & Val. Lower jaw longer than rest of head. South Atlantic coast of United States and southward, also on the Pacific coast southward.

Instead of the synonymy in the text read: (*Hemirhamphus roberti* Cuv. & Val., XIX, 1846, 24; Günther VI, 263, *Hemirhamphus unifasciatus* of most recent American authors, not of Ranzani, whose species is the short billed one.)

A discussion of the species of this genus is given by Meek & Goss, Proc. Ac. Nat. Sci. Phila., 1884.

[3] The species called in the Synopsis (p. 902), *Hemirhamphus brasiliensis*, should apparently stand as *Hemirhamphus pleei*.

[4] *Exocœtus mesogaster* Bloch, Ichthyol., XII, tab. 399 = *Exocœtus hillianus* Gosse. See Jordan & Gilbert, Proc. U. S. Nat. Mus., 1882, 588.)

[5] *Exocœtus obtusirostris* Günther, seems to be identical with *H. evolans*.

201.—EXOCŒTUS[1] Linnæus. '(185, 186)

673. **Exocœtus exiliens**[2] Gmelin. O. S. (613)
674. **Exocœtus rondeleti**[3] Cuv. & Val. S. O. Eu. (609)
675. **Exocœtus vinciguerræ**[4] Jordan & Meek. N. O. (609)
676. **Exocœtus volitans**[5] Linnæus. N. S. W. (611)
677. **Exocœtus heterurus** Rafinesque. N. S. Eu. (610, 613)
678. **Exocœtus furcatus** Mitchill. O. (612)
679. **Exocœtus californicus** Cooper. C. P. (608)
680. **Exocœtus gibbifrons** Cuv. & Val. O.

Order W.—LOPHOBRANCHII. (R.)

Family LXIX.—SYNGNATHIDÆ. (58, 59)

202.—SIPHOSTOMA Rafinesque (187)

681. **Siphostoma zatropis** Jordan & Gilbert. W. (618 b.)
682. **Siphostoma punctipinne** Gill. C. (618)
683. **Siphostoma californiense** Storer. C. (616)
684. **Siphostoma griseolineatum** Ayres. C. (616 b.)
685. **Siphostoma auliscus** Swain. C. (617 b.)
686. **Siphostoma barbaræ**[6] Swain & Meek. C. (616 c.)
687. **Siphostoma bairdianum**[7] Duméril. P.

---

[1] It is probable that *Cypselurus* is a young stage of *Exocœtus*. I have found on specimens of *Exocœtus mesogaster* two short barbels at the symphysis of the lower jaw, while in adult examples there is no trace of these appendages. For a full account of our species of this genus, see Jordan & Meek, Proc. U. S. Nat. Mus. 1885.

[2] The following is Gmelin's account of *Exocœtus exiliens*:
"*Exocœtus* pinnis ventralibus caudam attingentibus. D. 10, P. 15, V. 6, A. 11, C. 26. Habitat ad Carolinam, volitante statura simillimus, at vix digito longior, neque argenteus. Garden.
"Pinnæ pallidæ, fascia una alterave nigricante, ventrales * * apice pinnam caudæ attingentes, ¼ a caudæ remotæ, * * inter caput et anum mediæ, radio primo brevi, pectorales, radio primo et secundo brevibus; caudalis lobus inferior longior." (*Gmelin.*)

[3] *Exocœtus volador* Jordan, Proc. U. S. Nat. Mus., 1884, 34.

[4] *Exocœtus rondeletii*, Synopsis, p. 904, not of C. & V.; Lütken, Vid. Meddel. Naturh. Foren., 1876, 110.)

[5] *Exocœtus volitans* L. = *Exocœtus melanurus* Synopsis, p. 179; nec Cuv. & Val.; *Exocœtus exiliens* Synopsis, p. 904, not of Gmelin; *Exocœtus affinis* Günther, VI, 288; *Exocœtus roberti* Müller & Troschel, Schomburgk, Excurs. Barbadoes, 675 (probably).

[6] The species, called in the Synopsis, *Siphostoma bairdianum*, should stand as *Siphostoma barbaræ* Swain & Meek, Proc. U. S. Nat. Mus., 1884, 238. Santa Barbara.

[7] The original *Syngnathus bairdianus*, from the "coast of Mexico near California," proves to be a different species, having the technical characters of *S. affine*, but with the snout longer and the crest on top of head rather feebler. The following is Duméril's original description:
Head scarcely ⅙ of total length, a little longer than dorsal base; muzzle longer by a third than postocular part of head and equal to distance from front of eye to second ring; median crest of head and nape feeble; that of opercle very small. Rings 17+31. Tail at least half longer than trunk. Dorsal on 3+6 rings. P. 15, D. 30, A. 3, C. 6. Yellowish, sutures marked, except below, by a brown line. Coast of Mexico, near California.

688. Siphostoma leptorhynchum Girard. C. (617)
689. Siphostoma floridæ Jordan & Gilbert. S. (615 b.)
690. Siphostoma affine Günther. S. W. (614 b.)
691. Siphostoma louisianæ Günther. S. (615)
692. Siphostoma fuscum Storer. N. (614)
693. Siphostoma mackayi[1] Swain & Meek. W.
694. Siphostoma crinigerum[2] Bean & Dresel. S. W.

203.—DORYRHAMPHUS[3] Kaup.

695. Doryrhamphus californiensis Gill. P.

204.—HIPPOCAMPUS[4] Linnæus.

696. Hippocampus ingens Girard. C. P. (620)
697. Hippocampus punctulatus Guichenot. W. (619 b.)
698. Hippocampus hudsonius Dekay. N. S. (619 c.)
699. Hippocampus stylifer Jordan & Gilbert. S. (619 d.)
700. Hippocampus zosteræ Jordan & Gilbert. S. (619 e.)

Order X.—HEMIBRANCHII. (S)

Family LXX.—MACRORHAMPHOSIDÆ. (60)

205.—MACRORHAMPHOSUS[5] Lacépède. (189)

701. Macrorhamphosus scolopax Linnæus. Eu.. (621)

[1] *Siphostoma mackayi* Swain & Meek, Proc. U. S. Nat. Mus., 1884, 239; Key West. In this paper is a very useful analysis of the characters of the species of this genus, supplementary to a paper on the same subject by Mr. Swain, Proc. U. S. Nat. Mus., 1882, 307.

[2] *Siphostoma crinigerum* Bean & Dresel, Proc. Biol. Soc. Washington, II, 1884, 99. Swain & Meek, Proc. U. S. Nat. Mus., 1884, 239. Pensacola to Key West.

[3] DORYRHAMPHUS Kaup.

(Kaup, Lophobranchii, 1856, 54; type *Doryrhamphus excisus* Kaup.)
This genus differs from *Siphostoma* chiefly in the position of the egg-pouch of the male, which is under the abdomen instead of the tail. The angles of the body are strongly ridged. Tropical seas. (Δορυ, lance; ῥαμφος, snout.)

Doryrhamphus californiensis Gill.
Yellowish brown, with a black streak from snout to axil. Snout half as long as head, its crest formed of about ten irregular teeth, behind which are two others. Double frontal crest well serrated. Ridge under orbit unarmed, but on side of snout it is well serrated. Chin prominent but unarmed. Pectorals as long as opercle. Caudal as long as snout. D. 25. Rings 20+16. Cape San Lucas (*Gill*). The types are lost and no specimens have been since recorded.
(Gill, Proc. Ac. Nat. Sci. Phila., 1862, 284: *Doryichthys californiensis* Günther VIII, 186.)

[4] The family *Hippocampidæ* should be, apparently, reunited with the *Syngnathidæ*. I here omit *Hippocampus hippocampus* (= *heptagonus* Raf.; *antiquorum*, Leach), not believing that that species has been actually taken in American waters.

[5] The reasons for using the name *Macrorhamphosus* for this genus instead of *Centriscus* are stated in Proc. U. S. Nat. Mus., 1882, 575. The original type of *Centriscus* is C. *scutatus*.

A valuable discussion of "the mutual relations of the *Hemibranchiate* fishes" is given by Dr. Gill, Proc. Ac. Nat. Sci. Phila., 1884, 154.

Family LXXI.—FISTULARIIDÆ. (61)

206.—FISTULARIA Linnæus. (190)

702. Fistularia tabaccaria Linnæus. S. W. (622)
703. Fistularia serrata Cuvier. O. (623)
704. Fistularia depressa[1] Günther. P.

Family LXXII.—AULOSTOMIDÆ. (62)

207.—AULOSTOMA Lacépède. (191)

705. Aulostoma maculatum Valenciennes. W. (624)

Family LXXIII.—AULORHYNCHIDÆ. (63)

208.—AULORHYNCHUS Gill. (191)

706. Aulorhynchus flavidus Gill. C. A. (625)

Family LXXIV.—GASTEROSTEIDÆ. (64)

209.—PYGOSTEUS Brevoort.

707. Pygosteus pungitius Linnæus. N. Eu. (626)
707 b. *Pygosteus pungitius concinnus* Richardson. Vn.
707 c. *Pygosteus pungitius brachypoda* Bean. G.

210.—EUCALIA Jordan.

708. Eucalia inconstans Kirtland. Vn. (627)
708 b. *Eucalia inconstans cayuga* Jordan. Vne.

211.—GASTEROSTEUS Linnæus. (193)

709. Gasterosteus williamsoni[2] Girard. T.
710. Gasterosteus microcephalus Girard C. A. (628)
711. Gasterosteus (gymnurus?) cuvieri Girard. G. (629)
711 b. *Gasterosteus (cuvieri?) wheatlandi* Putnam. N.
712. Gasterosteus atkinsi Bean. Vne. (630)
713. Gasterosteus aculeatus Linnæus. N. Eu. (631)
713 b. *Gasterosteus aculeatus cataphractus* Pallas. A. (631 b)

212.—APELTES Dekay. (194)

714. Apeltes quadracus Mitchill. N. (632)

---

[1] *Fistularia depressa* Günther, Rept. Shore Fishes; Challenger, 1880, 69; East Indies, Australia, China, and Lower California. Abundant in the Gulf of California. Bones of the head less deeply sculptured than in *F. serrata*, but with the two upper lateral ridges of the snout also serrated; interorbital space nearly flat. Two middle ridges on upper surface of snout not very close together, diverging again on anterior half of length of snout, converging again finally on the foremost part. Body much depressed, nearly smooth, the skin being scarcely rough.

[2] For a description of this species, see Rosa Smith, Proc. U. S. Nat. Mus., 1883, 217. It is a true *Gasterosteus*, and not an *Eucalia*, although having the naked skin of the latter genus.

Order Y.—PERCESOCES.

Family LXXV.—MUGILIDÆ. (65)

213.—MUGIL Linnæus. (195)

715. Mugil cephalus[1] Linnæus. N. S. W. P. C. Eu. (633, 634)
716. Mugil gaimardianus[2] Poey. W.
717. Mugil curema[3] Cuvier & Valenciennes. N. S. W. P. (635)
718. Mugil trichodon[4] Poey. W.

214.—CHÆNOMUGIL[5] Gill.

719. Chænomugil proboscideus Günther. P.

215.—QUERIMANA[6] Jordan & Gilbert.

720. Querimana harengus Günther. P.
721. Querimana gyrans Jordan & Gilbert. S. W.

216.—AGONOSTOMUS[7] Bennett.

722. Agonostomus nasutus Günther. P.

[1] The American species (albula) seems to be identical with the European (cephalus). For a detailed account of the American Mugilidæ, see Jordan & Swain, Proc. U. S. Nat. Mus., 1884, 261.

[2] Mugil gaimardianus Poey, Ann. Lyc. Nat. Hist., N. Y., 1875, 64. Cuba, Key West. See Jordan & Swain, l. c.

[3] Mugil curema Cuv. & Val. = Mugil brasiliensis of authors, not of Agassiz. See Jordan & Swain, l. c.

[4] Mugil trichodon Poey. Cuba and Key West.

In the paper above cited, we have adopted the name Mugil brasiliensis for this species. This is perhaps too hasty, as the Mugil brasiliensis of Agassiz seems at least as likely to have been Mugil liza.

[5] CHÆNOMUGIL Gill.

(Gill, Proc. Ac. Nat. Sci., Phila., 1863, 169; type Mugil proboscideus Günther.)

Cleft of mouth lateral; lower jaw narrow; dentiform cilia in very many series, somewhat pavid; upper lip very thick; no adipose eyelid. Vertical fins scaly. One species known. ($Xανω$, to gape; Mugil.)

Chænomugil proboscideus Günther = Mugil proboscideus Günther, iii, 1861, 459. Mazatlan to Panama.

[6] QUERIMANA Jordan & Gilbert.

(Jordan & Gilbert, Proc. U. S. Nat. Mus., 1882, 588; type Myxus harengus Günther. This genus differs from Mugil chiefly in the presence of but two spines in the anal fin. The species are of small size, and some of them swim in schools at the surface.

Querimana harengus Günther. Myxus harengus Günther, iii, 467, 1861 = Querimana harengus Jordan & Swain, Proc. U. S. Nat. Mus., 1882, 274. Mazatlan to Peru; abundant.

Querimana gyrans Jordan & Gilbert, Proc. U. S. Nat. Mus., 1884, 26. Charleston to Key West.

[7] AGONOSTOMUS Bennett.

(Cestræus, Dajaus and Nestis Cuv. & Val.).

(Bennett, Proc. Comm. Zoöl. Soc., 1830, 166; type Agonostomus telfairi Bennett.)

Fresh water mullets with cleft of the mouth extending laterally about to front of eye. Small teeth in one or both jaws and sometimes on the vomer. Edge of lower lip rounded, not sharp. Stomach not gizzard like. Anal spines 3. Streams of mountainous regions in the tropics. ($Aγωνος$, not angulated; $στομα$, mouth.)

Agonostoma nasutum Günther, 111, 463; Jordan & Gilbert, Proc. U. S. Nat. Mus., 379. Streams of Lower California and Guatemala.

Family LXXVI.—ATHERINIDÆ. (66)

217.—ATHERINA Linnæus. (196)

723. Atherina eriarcha[1] Jordan & Gilbert. P.
724. Atherina carolina Cuv. & Val. S. (636)
725. Atherina stipes[2] Müller & Troschel. W. (637)
726. Atherina aræa[3] Jordan & Gilbert. W.

218.—LEURESTHES Jordan & Gilbert. (197)

727. Leuresthes tenuis Ayres. C. (638)

219.—LABIDESTHES Cope. (198)

728. Labidesthes sicculus Cope. Vc. (639)

220.—MENIDIA Bonaparte. (199)

729. Menidia laciniata Swain. S. (640)
730. Menidia vagrans Goode & Bean. S. (641)
731. Menidia notata Mitchill. N. (642)
732. Menidia audens Hay. Vs. (642b)
733. Menidia beryllina Cope. Vc. (643)
734. Menidia menidia[4] Linnæus. S. (644)
735. Menidia peninsulæ Goode & Bean. S. (645)

221.—ATHERINOPSIS Girard. (200)

736. Atherinopsis californiensis Girard. C. (646)

222.—ATHERINOPS Steindachner. (201)

737. Atherinops affinis Ayres. C. (647)

Family LXXVII.—SPHYRÆNIDÆ. (67)

223.—SPHYRÆNA Bloch. (202)

738. Sphyræna argentea Girard. C. P. (648)
739. Sphyræna borealis[5] De Kay. N. (649)
740. Sphyræna guaguanche Cuv. & Val. S. W. (650)
741. Sphyræna picuda Bloch & Schneider. S. W. (650 b.)
742. Sphyræna ensis Jordan & Gilbert. P.

---

[1] *Atherinella eriarcha* Jordan & Gilbert, Proc. U. S. Nat. Mus., 1881, 348. Mazatlan to Peru.

[2] *Atherina stipes* Müller & Troschel = *Atherina laticeps* Poey = *Atherina velieana* Goode & Bean. See Jordan & Gilbert, Proc. U. S. Nat. Mus., 1884, 116.

[3] *Atherina aræa* Jordan & Gilbert, Proc. U. S. Nat. Mus., 1884, 27. Key West.

[4] Called *Menidia bosci* in the Synopsis, pp. 408, 909.

[5] Called *Sphyræna spet* in the Synopsis, p. 411. Ours is, however, apparently distinct from the latter species, which is European.

[6] *Sphyræna ensis* Jordan & Gilbert, Bull. U. S. Fish Comm., 1881, 106, based on *Sphyræna forsteri* Steindachner, Ichth. Beiträge, VII, 4, 1878, not *Sphyræna forsteri* C. & V.

Body moderately elongate; eye 6 to 7 in head; snout 2¼; pectoral 2¾. Pectoral reaching about to front of first dorsal. Ventrals inserted before first dorsal. Canine teeth of lower jaw, palatines, and inner row of premaxillary very large, much as in *S. picuda*. Maxillary reaching about to front of dorsal. Silvery, darker above, with traces of numerous vague darker cross-bars. Head 4; depth 8 or 9. D. V-1, 9; A. 11. s. Lat. l. 110. Gulf of California to Panama.

For a detailed account of our species of this genus, see Meek & Newland, Proc. Ac. Nat. Sci. Phila., 1884.

Family LXXVIII.—POLYNEMIDÆ. (68)

224.—POLYNEMUS Linnæus.

743. **Polynemus virginicus**[1] Linnæus. W. (650c)
744. **Polynemus approximans**[2] Lay & Bennett. P.
745. **Polynemus opercularis**[3] Gill. P.
746. **Polynemus octonemus**[4] Girard. S.

ORDER Z.—PERCOMORPHI.[5]

Family LXXIX.—AMMODYTIDÆ. (69)

225.—AMMODYTES Linnæus. (204, 205)

747. **Ammodytes americanus** DeKay. N. (652, 656)
747b. *Ammodytes americanus personatus* Girard. A. C. (653)
748. **Ammodytes alascanus** Cope. A. (654)
749. **Ammodytes dubius** Reinhardt. B. (655)

Family LXXX.—ECHENEIDIDÆ. (70)

226.—ECHENEIS. (206)

750. **Echeneis naucrates** Linnæus. N. S. O. W. P. C. (657)

227.—PHTHEIRICHTHYS Gill. (206b.)

751. **Phtheirichthys lineatus** Menzies. S. W. (657 b.)

228.—REMORA Gill. (206c)

752. **Remora remora** Linnæus. S. O. W. P. C. (658)
753. **Remora brachyptera** Lowe. W. O. (659)
754. **Remora albescens**[6] Temminck & Schlegel. P. S.

229.—RHOMBOCHIRUS Gill. (207)

755. **Rhombochirus osteochir** Cuvier. O. W. (660)

---

[1] *Polynemus virginicus* L. Syst. Nat.=*Polydactylus plumieri* Lacépède. See Jordan, Proc. U. S. Nat. Mus., 1884, 118.

[2] *Polynemus approximans* Lay & Bennett, Beechey's Voyage, Zool. Fish, 57; Günther, Fish. Centr. Amer., 1869, 423. Gulf of California to Panama.

[3] *Trichidion opercularis* Gill, Proc. Ac. Nat. Sci. Phila., 1863, 169 = *Polynemus melanopoma* Günther, Fish. Centr. Amer. 1869, 421. Gulf of California to Panama.

[4] *Polynemus octofilis* Gill is without much doubt the adult form of *P. octonemus*. See Jordan & Gilbert, Proc. U. S. Nat. Mus., 1882, 590. The pectoral fin grows darker in color and the pectoral filaments shorter with age in other species of *Polynemus* and probably in this one also.

[5] *Percomorphi* and *Pharyngognathi* Cope, Trans. Am. Philos. Soc. Phila., 1871, 458 (exclusive of the *Rhegnopteri*=*Polynemidæ*, which have the ventral fins truly abdominal and may be placed in the *Percesoces*.)

[6] *Echeneis albescens* Temminck & Schlegel, Fauna Japonica, Poiss., 272; Günther II, 377; Streets, Bull. U. S. Nat. Mus., 1877, VII, 54. Coasts of Eastern Asia, a specimen taken at La Paz, Gulf of California (*Streets*) and in the Gulf of Mexico (*Bean*). D. XIII-22; A. 22.

The *Echeneididæ* are regarded by Dr. Gill as constituting a distinct suborder, *Discocephali*, defined by him Proc. U. S. Nat. Mus., 1882, 563.

Family LXXXI.—ELACATIDÆ. (71)

230.—ELACATE Cuvier. (208)

756. **Elacate canada** Linnæus. S. W. O. (661)

Family LXXXII.—XIPHIIDÆ. (72)

231.—XIPHIAS Linnæus. (209)

757. **Xiphias gladius** Linnæus. O. N. S. W. C. (662)

232.—TETRAPTURUS Rafinesque. (210)

758. **Tetrapturus albidus** Poey. W. S. (663)

233.—ISTIOPHORUS Lacépède. (211)

759. **Istiophorus americanus**[1] Cuv. & Val. (665)

Family LXXXIII.—TRICHIURIDÆ. (73)

234.—TRICHIURUS Linnæus. (212)

760. **Trichiurus lepturus** Linnæus. O. S. W. P. (666)

235.—BENTHODESMUS Goode & Bean. (212b.)

761. **Benthodesmus elongatus** Clarke. B. (666b.)

236.—LEPIDOPUS Gouan.

762. **Lepidopus caudatus** Euphrasen. O. P.

---

[1] The genuine *Istiophorus gladius* is an East Indian species, not known from our coasts. The American species is:

Istiophorus americanus Cuv. & Val. *Sail-fish; Spike-fish.* Bluish-black, paler below; dorsal dusky-bluish; its membranes with many nearly round black spots, from $\frac{1}{8}$ to $\frac{1}{4}$ diameter of orbit. Snout, from eye, $2\frac{1}{4}$ times length of rest of head. Lower jaw $2\frac{1}{4}$ in head. Front of eye nearly midway between tip of lower jaw and edge of opercle. Interorbital space broad, flattish, $1\frac{2}{3}$ in postorbital part of head. Maxillary reaching to slightly beyond eye, which is $3\frac{1}{4}$ in postorbital part of head and 10 in snout. Sword narrow, regularly tapering, depressed, its upper and lower surfaces both rounded, its edges blunt and rougher than its upper side. For its entire length it is nearly twice as broad as deep. Breadth of snout at the middle point between its tip and the eye contained 25 times in its length from the eye. Longest dorsal spine $\frac{4}{3}$ total length of head. Ventrals $1\frac{5}{6}$ in head. Pectorals $3\frac{3}{7}$. Caudal lobes $1\frac{3}{4}$. D. XLI-7; A. 9-7. Head $2\frac{2}{3}$ ($3\frac{1}{4}$ in length with caudal); depth about 6. Length of specimen described (Key West) 6 feet.

West Indies and warmer parts of the Atlantic, north to Cape Cod and France. Differing from the East Indian *I. gladius* in the longer and slenderer sword and in the shorter dorsal fin.

(? *Makaira nigricans* Lacépède, Hist. Nat. Poiss. IV, 688, 1803. *Histiophorus americanus* Cuv. & Val., VIII, 303, 1831; ? *Histiophorus gracilirostris* C. & V., VIII, 308; ? *Histiophorus ancipitirostris* Cuv. & Val., VIII, 309. I here restore the original orthography of the name *Istiophorus*.)

[2] LEPIDOPUS Gouan.

(Gouan, Hist. Poiss. 1770, 185; type *Lepidopus gouani* Bl. & Schn. = *Trichiurus caudatus* Euphrasen.)

This genus differs from *Trichiurus* chiefly in the less elongate form of the tail, which

Family LXXXIV.—SCOMBRIDÆ. (74)

237.—SCOMBER Linnæus. (213)
§ *Pneumatophorus* Jordan & Gilbert.
763. Scomber colias[1] Gmelin. Eu. N. S. P. C. (667, 667 b.)
§ *Scomber*.
764. Scomber scombrus Linnæus. N. S. O. Eu. (668)

238.—AUXIS Cuvier. (214)
765. Auxis thazard Lacépède. W. N. (Acc.) O. (669)

239.—SCOMBEROMORUS Lacépède. (215)
766. Scomberomorus concolor Lockington. C. (670)
767. Scomberomorus maculatus Mitchill. N. S. P. (671)
768. Scomberomorus regalis Bloch. W. (672)
769. Scomberomorus cavalla[2] Cuvier. W. S. (673)

240.—ACANTHOCYBIUM[3] Gill.
770. Acanthocybium solandri Cuv. & Val. W. O.

is provided with a small, deeply forked caudal fin. The ventral fins are represented by a pair of scale-like appendages. A single species; pelagic. ($λεπίς$, scale; $πούς$, foot.)

Lepidopus caudatus. Scabbard-fish. For description, see Günther II, 344. Pelagic; a specimen taken by John Xantus at Cape St. Lucas.

[1] It is probable that *Scomber pneumatophorus* is identical with *Scomber colias*.

[2] This species was first indicated as *Cybium caralla* Cuvier, Régne Animal, 1829. It is the king-fish of the Florida Keys, a food fish of the highest importance. For a detailed account of the species of *Scomberomorus* see Meek and Newland, Proc. Ac. Nat. Sci. Phila., 1884.

[3] ACANTHOCYBIUM Gill.

(Gill, Proc. Ac. Nat. Sci. Phila., 1862; type *Cybium sara* Bennett.)

This genus is allied to *Scomberomorus*, but shows several of the peculiarities of the sword-fishes, indicating a transition toward the *Xiphiidæ*. The head is very long, slender, and pointed, the mandible being longer than the upper jaw, the jaws forming a sort of beak; cleft of the mouth extending to below the eye; the posterior part of the maxillary covered by the preorbital; both jaws armed with a close series of trenchant teeth, ovate or truncate; their edges finely serrate; villiform teeth on vomer and palatines; gills formed as in *Xiphias*, their laminæ forming a net-work; scales small, scarcely forming a corselet; those along the base of dorsal enlarged and lanceolate; keel strong; caudal spinous dorsal very long, its spines about 25 in number.

Very large mackerels, pelagic; probably a single species widely distributed; most abundant about the Florida Straits. ($Ἄκανθα$, spine; *Cybium*.)

Acanthocybium solandri. *Peto*; *Wahoo*; *Barracotta*.

Iron gray, dark above; paler below; no distinct markings; fins colored like the body; eye 5 in snout; gape more than half length of head; premaxillaries in front prolonged in a sort of beak which is nearly half length of snout; teeth somewhat irregular, the posterior much largest. Dorsal spine mostly subequal, the highest, behind the middle of the fin, 5⅔ in head; dorsal and anal lobes low. Caudal lobes short, very abruptly spreading, their length about ⅜ head. Pectoral not quite half head. D. XXIV-1, 12–IX; A. 1, 12–IX. Length 4 to 8 feet. Tropical seas; not rare about Cuba, where it spawns; north to Key West.

(*Cybium solandri* Cuv. & Val., VIII, 1831, 192; *Cybium sara* Bennett, Beechey's Voyage, Zoölogy, 1849, 63; *Cybium sara* Günther, II, 373; *Cybium petus* Poey, Memorias Cuba, II, 234, 1860; *Acanthocybium petus* Poey, Enum. Pisc. Cubens., 1875, 73. Lütken, Spolia Atlantica, 1880, 481–597; *Cybium veranyi* Doderlein, Giorn. Sci. Natur. Econ. Palermo, 1872.

241.—SARDA Cuvier. (216)
771. Sarda sarda Bloch. Eu. N. (674)
772. Sarda chilensis Cuv. & Val. C. P. (675)

242.—ORCYNUS Cuvier. (217)
773. Orcynus alalonga Gmelin. Eu. S. C. O. (676)
774. Orcyuus thynnus Linnaeus. Eu. S. N. O. (677)

243.—EUTHYNNUS Lütken. (218)
775. Euthynnus alliteratus Rafinesque. S. W. Eu. (678)
776. Euthynnus pelamys Linnaeus. Eu. S. O. (679)

Family LXXXV.—CARANGIDÆ.[1] (75)
244.—DECAPTERUS Bleeker. (220)
777. Decapterus punctatus Agassiz. S. W. (682)

---

[1] The following analysis of genera of *Carangidæ* may be substituted for that given in the synopsis:

a. Premaxillaries protractile.
  b. Pectoral fins long, falcate; anal similar to soft dorsal, its base longer than abdomen; maxillary with a supplemental bone. (*Caranginæ*.)
    c. Dorsal outline more strongly curved than ventral outline.
      d. Dorsal and anal each with a single detached finlet; body slender. DECAPTERUS.
      dd. Dorsal and anal without finlets.
        e. Lateral line with well-developed scutes for its entire length; body elongate............TRACHURUS.
        ee. Lateral line with scutes on its straight posterior portion only (these sometimes very few and small, especially in those species with the body much compressed).
          f. Shoulder girdle with a deep cross-furrow at its junction with the isthmus, above which is a fleshy projection; body elongate............TRACHUROPS.
          ff. Shoulder girdle normal; its surface even; body deeper.
            g. Body oblong or more or less elevated, not as below............CARANX.
            gg. Body broad-ovate, very strongly compressed, its outlines everywhere trenchant, the anterior profile nearly vertical; scutes almost obsolete............VOMER.
        eee. Lateral line without any scutes; body short and elevated, strongly compressed............SELENE.
    cc. Dorsal outline less strongly curved than ventral; body much compressed, its outlines everywhere trenchant; armature of lateral line obsolete or nearly so.
                                          CHLOROSCOMBRUS.
  bb. Pectoral fin short, not falcate.
    h. Maxillary without supplemental bone; anal fin similar to soft dorsal, its base much longer than abdomen; tail unarmed. (*Trachynotinæ*.)
      d. Forehead convex; teeth small or deciduous............TRACHYNOTUS.
    hh. Maxillary with a distinct supplemental bone; anal fin shorter than soft dorsal, its base not longer than abdomen. (*Seriolinæ*.)
      i. Dorsal spines low and weak; pectoral fins short.
        j. Dorsal and anal fins without finlets.
          k. Membrane of dorsal spines disappearing with age. NAUCRATES.
          kk. Membrane of dorsal spines persistent............SERIOLA.
        jj. Dorsal and anal fins each with a detached two-rayed finlet.
                                          ELAGATIS.
      ii. Dorsal spines strong, ending in very long filaments; pectoral fins elongate............NEMATISTIUS.

778. **Decapterus macarellus** Cuv. & Val. W. S. (683)
778 b. *Decapterus macarellus hypodus*[1] Gill. P.

245.—TRACHURUS Rafinesque. (219)

779. **Trachurus picturatus** Bowdich. C. Eu. P. (680)
780. **Trachurus trachurus** Linnaeus. W. P. (681)

246. TRACHUROPS Gill.

781. **Trachurops crumenophthalmus** Bloch. W. P. (684)

247.—CARANX Lacépède.

§ *Hemicaranx* Bleeker.

782. **Caranx amblyrhynchus** Cuv. & Val. S. W. (689)

§ *Uraspis* Bleeker.

783. **Caranx vinctus**[2] Jordan & Gilbert P.
784. **Caranx bartholomæi**[3] Cuv. & Val. W. (687, 688)

§ *Caranx.*

785. **Caranx chrysus** Mitchill. N. S. W. (685)
785 b. *Caranx chrysus caballus* Günther. P. W. (686)
786. **Caranx latus**[4] Agassiz. S. W. P. (690)
787. **Caranx hippos** Linnaeus. N. S. W. P. (691)

§ *Gnathanodon* Bleeker.

788. **Caranx speciosus**[5] Forskål. P.

§ *Citula* Cuvier.

789. **Caranx dorsalis**[6] Gill. P.

§ *Blepharis* Cuvier.

790. **Caranx crinitus** Mitchill. N. S. W. P. (692)

aa. Premaxillaries not protractile (except in the very young); pectoral fins short rounded; soft dorsal similar to anal, both much longer than abdomen; lateral line unarmed. (*Scombroidinæ.*)

l. Maxillary without supplemental bone; no pterygoid teeth; scales linear, imbedded .... OLIGOPLITES.

A detailed account of the American species of *Caranginæ* is given by Jordan & Gilbert, Proc. U. S. Nat. Mus., 1883, 188.

[1] *Decapterus hypodus* Gill, Proc. Ac. Nat. Sci., Phila., 1862, 261; Jordan & Gilbert, Proc. U. S. Nat. Mus., 1882, 352; 1883, 190. Cape San Lucas.

[2] *Caranx vinctus* Jordan & Gilbert, Proc. U. S. Nat. Mus., 1882, 349. Mazatlan.

[3] *Caranx bartholomæi* Cuv. & Val., IX, 1833, 100 = *Caranx cibi* Poey, Memorias Cuba, II, 221, 1860 = *Caranx beani* Jordan, Proc. U. S. Nat. Mus., 1880, 486. See Jordan & Gilbert, Proc. U. S. Nat. Mus., 1884, 32.

[4] *Caranx latus* Agassiz; *Caranx fallax* Cuv. & Val. See Jordan & Gilbert, Proc. U. S. Nat. Mus., 1883, 200.

[5] *Scomber speciosus* Forskål, Descr. Anim., 1775, 54 = *Caranx panamensis* Gill, Proc. Ac. Nat. Sci. Phila., 1863, 166. See Jordan & Gilbert, Proc. U. S. Nat. Mus., 1883, 201. Mazatlan to Panama and west to the Red Sea.

[6] *Carangoides dorsalis* Gill, Proc. U. S. Nat. Mus., 1863, 166 = *Caranx otrynter* Jordan & Gilbert, Proc. U. S. Nat. Mus., 1883, 202. Mazatlan to Panama.

248.—VOMER Cuvier.

791. **Vomer setipinnis** Mitchill. N. S. W. P. (694)

249.—SELENE Lacépède. (223)

792. **Selene œrstedi**[1] Lütken. P.
793. **Selene vomer** Linnæus. N. S. W. P. (693)

250.—CHLOROSCOMBRUS Girard. (224)

794. **Chloroscombrus chrysurus** Linnæus. S. W. (695)
795. **Chloroscombrus orqueta**[2] Jordan & Gilbert. P.

251.—TRACHYNOTUS Lacépède.

796. **Trachynotus carolinus** Linnæus. N. S. W. P. ? (696)
797. **Trachynotus argenteus**[3] Cuv. & Val. N.
798. **Trachynotus rhodopus**[4] Gill. W. P. (698)
799. **Trachynotus kennedyi**[5] Steindachner. P.
800. **Trachynotus rhomboides** Bloch. S. W. (697)
801. **Trachynotus glaucus** Bloch. S. W. (699)
802. **Trachynotus fasciatus**[6] Gill. P.

252.—NAUCRATES Rafinesque. (226)

803. **Naucrates ductor** Linnæus. O. (700.)

253.—SERIOLA Cuvier. (227)

804. **Seriola zonata** Mitchill. N. (704)
804b. *Seriola zonata carolinensis* Holbrook. S. (703)
805. **Seriola dumerili**[7] Risso. S. W. Eu.
805b. *Seriola dumerili lalandi.* S. W. (701b.)

---

[1] *Selene orstedi* Lütken, Spolia Atlantica, 1880, 144; Jordan & Gilbert, l. c. 205. Mazatlan to Panama.

[2] *Chloroscombrus orqueta* Jordan & Gilbert, Proc. U. S. Nat. Mus., 1882, 646. Magdalena Bay to Panama.

[3] *Trachynotus argenteus* Cuv. & Val., VIII, 413. According to Dr. Bean, this is probably a valid species, allied to *T. carolinus*, but with the body deeper, the depth being half the length without caudal. New York.
 A review of the American species of *Trachynotus* is given by Meek and Goss in the Proc. Ac. Nat. Sci. Phila., 1884.

[4] The species called in the synopsis "*Trachynotus goreensis*" should stand as *Trachynotus rhodopus* Gill. *Permit. Palometa.* West Indies, north to Florida and Lower California. Instead of the synonymy in the synopsis read: *Trachynotus rhodopus* (young) and *T. nasutus* (very young) Gill, Proc. Ac. Nat. Sci. Phila, 1863, 85; *Trachynotus goreensis* Günther, II, 483, in part, not of Cuv. & Val.; *Trachynotus goreensis* of recent American writers; *Trachynotus carolinus* Poey, Enum. Pisc. Cubens., 86.
 This species reaches a larger size than the others in our waters. It has fewer fin rays than *T. carolinus*, and young and old are much more elongate than in *T. rhomboides* or than in the African *T. goreensis*.

[5] *Trachynotus kennedyi* Steindachner, Ichth. Beitr., VI, 47. Mazatlan to Panama.

[6] *Trachynotus fasciatus* Gill, Proc. Ac. Nat. Sci. Phila. 1863, 86. Mazatlan to Panama.

[7] *Seriola dumérili* Risso. *Amber Jack.*
 Grayish; silvery below; a gilt band through eye to base of caudal; another through temporal region to front of soft dorsal; no dark cross-bands; fins plain. Very close to *S. lalandi*, but reaching a smaller size, and with the body deeper and little com-

806. Seriola mazatlana¹ Steindachner. P.
807. Seriola dorsalis Gill. C. P. (701)
808. Seriola fasciata Bloch. S. (705)
809. Seriola rivoliana Cuv. & Val. S. W. Eu. (702, 702 b.)

254.—ELAGATIS Bennett. (228)

810. Elagatis pinnulatus Poey. W. (706)

255.—NEMATISTIUS² Gill.

811. Nematistius pectoralis Gill. P.

256.—OLIGOPLITES Gill. (229)

812. Oligoplites altus³ Günther. P.
813. Oligoplites saurus Bloch & Schneider. S. W. P. (707)

Family LXXXVI.—POMATOMIDÆ. (76)

257.—POMATOMUS Lacépède. (230)

814. Pomatomus saltatrix Linnæus. N. S. W. Eu. O. (708)

Family LXXXVII.—NOMEIDÆ. (76 b.)

258.—NOMEUS Cuvier. (231)

815. Nomeus gronovii Gmelin. W. O. (709)

Family LXXXVIII.—STROMATEIDÆ. (77)

259.—STROMATEUS Linnæus. (232)

§ *Rhombus* Lacépède.

816. Stromateus paru Linnæus. S. W. (710)

---

pressed; mouth larger than in *S. dorsalis*, about as in *S. lalandi*, the maxillary reaching middle of pupil, 2⅕ in head. Lobes of dorsal and anal low, not quite half length of head. Nape scarcely carinated. Head 3⅐; depth 3. D. VII-I, 32; A. II-I, 21; L. 24 inches. Mediterranean to West Indies, north to Key West and Pensacola.

(*Trachurus aliciolus* Rafinesque Caratteri, etc., 1810, 42; *Trachurus fasciatus* Rafinesque, Indice d'Ittiologia Sicil., 1810, 21; *Caranx dumérili* Risso, Ichthyologie Nice, 1810, 175; *Seriola dumérili* Cuv. & Val., IX, 201, 1833; Günther, II, 462; ? *Seriola semicoronata* Poey, Memorias Cuba, II, 1860, 232.)

An analysis of the characters of the species of *Seriola* is given by me in Proc. U. S. Nat. Mus., 1884, 123. A more recent (unpublished) study of these fishes by Mr. Rufus L. Green indicates the probable identity of *S. lalandi* with *S. aliciola* (*dumérili*), *S. falcata* with *S. rivoliana*, and (probably) *S. mazatlana* with *S. dorsalis*.

¹ *Seriola mazatlana* Steindachner, Ichth. Beiträge, V. 8, 1876. Mazatlan.

² NEMATISTIUS Gill.

(Gill, Proc. Ac. Nat. Sci. Phila., 1862, 258; type, *Nematistius pectoralis* Gill).

This genus differs from *Seriola* chiefly in the development of the spinous dorsal and pectoral fins, the former being composed of eight very long filamentous spines, the latter being acuminate and nearly twice as long as the ventrals. The lateral line is nearly straight and is not keeled on the caudal peduncle. Ventral rays, I, 5, the inner ray much branched to the base. One species known. Large fishes of an imposing appearance.

Nematistius pectoralis Gill, l. c. Gulf of California to Panama; not rare.

³ *Chorinemus altus* Günther, Fishes Centr. Amer., 1869, 433. Mazatlan to Panama.

[73] CATALOGUE OF THE FISHES OF NORTH AMERICA.

§ *Stromateus.*

817. **Stromateus medius**[1] Peters. P.
818. **Stromateus simillimus** Ayres. C. (711)

§ *Poronotus.*

819. **Stromateus triacanthus** Peck. N. (712)

260.—**LEIRUS** Lowe. (233)

820. **Leirus perciformis** Mitchill. N. (713)

Family LXXXIX.—LAMPRIDIDÆ. (78)

261.—**LAMPRIS** Retzius. (234)

821. **Lampris guttatus** Brünnich. O. (714)

Family XC.—CORYPHÆNIDÆ. (79)

262.—**CORYPHÆNA** Linnæus. (235.)

822. **Coryphæna hippurus**[2] Linnæus. O. S. W. (715, 716)

Family XCI.—BRAMIDÆ. (80)

263.—**PTERACLIS** Gronow. (236)

823. **Pteraclis carolinus** Cuv. & Val. O. (717)

264.—**BRAMA** Bloch & Schneider. (236 b.)

824. **Brama raji** Bloch. C. N. En. O. (717 b.)

Family XCII.—ICOSTEIDÆ.[3] (101)

265.—**ICOSTEUS** Lockington. (332)

825. **Icosteus ænigmaticus** Lockington. B. C. (969)

266.—**ICICHTHYS** Jordan & Gilbert. (333)

826. **Icichthys lockingtoni** Jordan & Gilbert. B. C. (970)

---

[1] *Stromateus medius* Peters, Berliner Monatsberichte, 1869, 707; Jordan, Proc. Ac. Nat. Sci. Phila., 1883, 284.

[2] *Coryphæna equisetis* has not been authentically recorded from our coasts. It may, therefore, be omitted. The common Dolphin or Dorado of our South Atlantic and Gulf coasts is *Coryphæna hippurus* L.
This species is in life of a very bright greenish olive, with small round blue spots. The top of the head in the males is much elevated, forming a high sharp crest. Head 4¾; depth 5; ventral inserted slightly behind upper ray of pectoral, its length 1¼ in in head; pectoral 1¼. D. 59 to 63; A. 29. Pelagic, north on our coast to Cape Cod; very abundant from South Carolina to Texas. L. 3 to 5 feet. The specific names *punctulata, globiceps, sueuri, dorado, guttata,* and *punctata* all belong to this species.

[3] The position of our family ICOSTEIDÆ is near or under the family BRAMIDÆ, as has been shown by Dr. Steindachner, Ichth. Beitr. XII, 22. The genus *Bathymaster* is apparently not a natural ally of *Icosteus.*

Family XCIII.—ZENIDÆ. (81)

267.—ZENOPSIS Gill. (237)

827. Zenopsis ocellatus Storer. B. (718)

Family XCIV.—BERYCIDÆ. (82)

268.—STEPHANOBERYX[1] Gill.

828. Stephanoberyx monæ Gill. B.

269.—CAULOLEPIS[2] Gill.

829. Caulolepis longidens Gill. B.

270.—PLECTROMUS[3] Gill.

830. Plectromus suborbitalis Gill. B.
831. Plectromus crassiceps Bean. B.

---

[1] STEPHANOBERYX Gill.

(Gill, Proc. U. S. Nat. Mus., 1883, 258; type *Stephanoberyx monæ* Gill.

"Berycids with an elongated claviform contour, body covered with cycloid scales; scarcely imbricated, and armed about the center with one or two erect spines; an oblong head, with a moderate convex snout and with thin osseous ridges, especially an inner V-shaped one on the crown, whose limbs diverge on each side of nape, and an outer sigmoid, one on each side, above the eyes, and continuous with one projecting from the nasal; the inner and outer ridges connected by a cross-bar on a line with the anterior margin of the orbit; rather small eyes, in the anterior half of the head, and the teeth small, acute, and in a band on the premaxillaries and dentaries (palate toothless), and with ventrals having one spine and five rays. Closely allied to *Melamphaës*." *Deep sea.* (Στεφανοϛ, crown; βῆρυξ, beryx.)

Stephanoberyx monæ Gill. Gulf stream, latitude 41°. (Gill, l. c. 258.)

[2] CAULOLEPIS Gill.

(Gill, Proc. U. S. Nat. Mus., 1883, 258; type *Caulolepis longidens* Gill.)

"Berycids with a laterally oval or broad pyriform contour; a compressed body, covered with small, pedunculated, leaf-like scales; an abruptly declivous forehead; small eyes; a pair of very long pointed teeth in front of upper jaw, closing in front of lower; a similar pair of still longer teeth in the lower, received in foveæ of the palate; on the sides of each jaw two long teeth, terminating in bulbous tips; a row of minute teeth on the posterior half of the maxillaries. Closely allied to *Anoplogaster*." *Deep sea.* (Καυλοϛ, stem; λεπιϛ, scale.)

Caulolepis longidens Gill. Deep sea; latitude 39°. (Gill, l. c. 258.)

[3] PLECTROMUS Gill.

(Gill, Proc. U. S. Nat. Mus., 1883, 257; type *Plectromus suborbitalis* Gill.)

"Berycids with an elongated form; moderate cycloid scales; an oblong head with a much decurved or truncate snout; rather small eyes, and teeth small, acute and in two rows in each jaw, of which those of the minor row, at least in the lower jaw, are largest, and palate toothless." Deep sea. (Πλῆκτρον, spur; ωμοϛ, shoulder); "two spines, one on each side of the nape, springing forward from the shoulder bones, give a strange appearance to the fish.")

Plectromus suborbitalis Gill. Gulf Stream, latitude 39°. (Gill, l. c., 257.)

Plectromus crassiceps Bean. Proc. U. S. Nat. Mus., 1885, 73. Gulf Stream.

271.—POROMITRA[1] Goode & Bean.

832. **Poromitra capito** Goode & Bean. B.

272.—HOPLOSTETHUS Cuv. & Val. (238)

833. **Hoplostethus mediterraneus** Cuv. & Val. B. Eu. (719)

Family XCV.—HOLOCENTRIDÆ.[2]

273.—HOLOCENTRUM Bloch. (239)

834. **Holocentrum ascensione**[3] Osbeck. W. (720)
835. **Holocentrum suborbitale**[4] Gill. P.

274.—MYRIPRISTIS[5] Cuv.

836. **Myripristis occidentalis** Gill. P.
837. **Myripristis pœcilopus** Gill. P.

---

[1] POROMITRA Goode & Bean.

(Goode & Bean, Bull. Mus. Comp. Zoöl., 1882, 215; type, *Poromitra capito* G. & B.).
Body short, compressed, scopeliform, covered with thin cycloid scales. Head very large (in young specimens nearly as long as trunk), its sides scaly. No barbel. Mouth very large, the lower jaw projecting. Margin of upper jaw composed of a long maxillary and a short premaxillary. Teeth very small, cardiform, on premaxillaries and lower jaw only. Opercula complete. Dorsal fin in middle of body, its origin not far behind ventrals, its spinous and soft portions about equal in length; anal much shorter than dorsal; the last rays of dorsal nearly above its middle. Pseudobranchiæ present. Gill openings very wide. Deep seas. ($\Pi o \rho o \varsigma$, pore; $\mu \iota \tau \rho a$, stomacher.)

Poromitra capito Goode & Bean.
Eye large, as long as snout; maxillary $3\frac{1}{4}$ in head. Scales as large as pupil, with concentric striæ. Insertion of dorsal midway between tip of snout and base of caudal; base of anal half that of dorsal; pectoral inserted low, its length twice its distance from the snout; ventrals minute, in advance of pectorals. Caudal (mutilated in the known specimens). Head $2\frac{1}{4}$ (in young). D. VII or VIII, 9; A. 9; V. 7 or 8; P. 12. Gulf Stream in lat. 34°. (*Goode & Bean.*)

(Goode & Bean, l. c., 214, 1882).

[2] The genera *Holocentrum* and *Myripristis*, shore fishes with long spinous dorsal, should probably be regarded as forming a family distinct from the *Berycidæ*, which are deep-sea fishes with a single dorsal, provided with but few spines, or even with none.

[3] This species, called in the text *Holocentrum pentacanthum*, should apparently stand as *Holocentrum ascensione* (Osbeck). In life, an oblique white bar descends backward from the eye; this disappears entirely in spirits. To the synonymy, add: (*Perca ascensionis* Osbeck, Iter Chin., 1771, 388; *Perca ascensionis* Gmelin, Syst. Nat., 1788, 1318; *Amphiprion matejuelo* Bloch & Schneider, Ichthyol., 1801, 206; *Holocentrum matejuelo* Poey, Memorias Cuba, II, 155, 1860.)

[4] *Holocentrum suborbitale* Gill, Proc. Ac. Nat. Sci. Phila., 1863, 86. Mazatlan to Panama. Abundant in rock-pools.

[5] MYRIPRISTIS Cuv.

(Cuvier, Règne Animal; type *Myripristis jacobus* Cuv. & Val.)
This genus is very closely related to *Holocentrum*, differing externally, chiefly in the absence of the large spine at the angle of the preopercle. The air-bladder is divided into two parts by a transverse constriction, and the pyloric cœca are rather

Family XCVI.—APHREDODERIDÆ. (83)

275.—APHREDODERUS Le Sueur. (240)

838. Aphredoderus sayanus Gilliams. (721)

Family XCVII.—ELASSOMIDÆ. (83b)

276.—ELASSOMA Jordan (722)

839. Elassoma zonatum Jordan. Vs. (722)
840. Elassoma evergladei[1] Jordan. Vsc.

Family XCVIII.—CENTRARCHIDÆ. (84)

277.—CENTRARCHUS Cuv. & Val. (242)

841. Centrarchus macropterus Lacépède. Vs. (723)

278.—POMOXYS Rafinesque. (243)

842. Pomoxys annularis Rafinesque. V. (724)
843. Pomoxys sparoides Lacépède. V. (725)

279.—ARCHOPLITES Gill. (244)

844. Archoplites interruptus Girard. T. (726)

280.—AMBLOPLITES Rafinesque. (245)

845. Ambloplites rupestris Rafinesque. V. (727)

281.—CHÆNOBRYTTUS Gill. (246)

846. Chænobryttus gulosus Cuv. & Val. V. (729)
846b. Chænobryttus gulosus antistius McKay. Vn. (728)

282.—ACANTHARCHUS Gill. (247)

847. Acantharchus pomotis Baird. Ve. (736)

283.—ENNEACANTHUS Gill.

848. Enneacanthus eriarchus Jordan. Vn. (731)
849. Enneacanthus obesus Baird. Ve. (732)
850. Enneacanthus gloriosus Holbrook. Vsc. (733)
851. Enneacanthus simulans Cope. Ve. (734)
851b. Enneacanthus simulans pinniger Gill & Jordan. Vsc.

284.—MESOGONISTIUS Gill.

852. Mesogonistius chætodon Baird. Ve. (735)

---

few (9). Species numerous in the tropical seas; gay-colored inhabitants of reefs and rock-pools.

Myriopristis occidentalis Gill, Proc. Ac. Nat. Sci. Phila., 1863, 87 = *Rhamphoberyx leucopus* Gill, l. c., 88. Gulf of California to Panama.

Myriopristis pœcilopus Gill. *Rhamphoberyx pœcilopus* Gill, l. c., 87; see Jordan & Gilbert, Proc. U. S. Nat. Mus., 1882, 364. Cape San Lucas; perhaps identical with the preceding.

[1] *Elassoma evergladei* Jordan, Proc. U. S. Nat. Mus., 1884, 323. Indian, Saint John's and Suwannee Rivers, Florida.

285.—LEPOMIS Rafinesque. (250)

§ *Apomotis* Rafinesque.

853. Lepomis cyanellus Rafinesque. V. (736)
854. Lepomis symmetricus Forbes. Vs. (737)
855. Lepomis phenax Cope & Jordan. Vc. (738)

§ *Lepomis.*

856. Lepomis ischyrus Jordan & Nelson. Vnw. (739)
857. Lepomis macrochirus Rafinesque. Vw. (740)
858. Lepomis mystacalis Cope. Vse. (741)
859. Lepomis elongatus Holbrook. Vse. (742)
860. Lepomis murinus Girard. Vsw. (743)
861. Lepomis punctatus Cuv. & Val. Vse. (744)
862. Lepomis miniatus Jordan. Vs. (745)
863. Lepomis auritus Linnæus. Vc. (746)
864. Lepomis megalotis[1] Rafinesque. Vw. (747, 749)
865. Lepomis garmani Forbes. Vw.
866. Lepomis marginatus Holbrook. Vse. (748)
867. Lepomis aquilensis[2] Baird & Girard. Vsw.
868. Lepomis humilis Girard. Vsw. (750)
869. Lepomis pallidus Mitchill. V. (751)

§ *Xystroplites* Jordan.

870. Lepomis heros Baird & Girard. Vsw. (752)
871. Lepomis euryorus McKay. Vn. (753)
872. Lepomis albulus Girard. Vsw. (754)

§ *Eupomotis* Gill & Jordan.

873. Lepomis holbrooki Cuv. & Val. Vse. (755)
674. Lepomis notatus Agassiz. Vs. (756)
875. Lepomis gibbosus Linnæus.[3] Vnc. (757)

286.—MICROPTERUS Lacépède. (251)

876. Micropterus salmoides Lacépède. V. (759)
877. Micropterus dolomiei Lacépède. V. (760.)

Family XCIX.—PERCIDÆ. (85)

287.—AMMOCRYPTA Jordan. (252)

878. Ammocrypta beani Jordan. Vs. (761)
879. Ammocrypta clara[4] Jordan & Meek. Vw.
880. Ammocrypta pellucida Baird. Vw. (762)
881. Ammocrypta vivax Hay. Vsw. (762 b.)

---

[1] *Lepomis bombifrons* is omitted, as being probably based on a form of *L. megalotis*.

[2] *Lepomis aquilensis* (*Pomotis aquilensis* Baird & Girard, Proc. Ac. Nat. Sci. Phila. 1854, 24), placed in the Synopsis as a synonym of *L. pallidus*, is a valid species. It is closely related to *L. megalotis*, but has much higher spines, and a long and very narrow opercular flap; a dusky patch on base of last rays of dorsal.

[3] *Lepomis lirus* McKay=*Pomotis pallidus* Agassiz is here omitted. Agassiz's very poor description applies well enough to *Chænobryttus gulosus*.

[4] *Ammocrypta clara* Jordan & Meek, Proc. U. S. Nat. Mus., 1884. Des Moines R., Iowa, and Red R., Arkansas.

288.—CRYSTALLARIA[1] Jordan & Gilbert.

882. Crystallaria asprella Jordan. Vs. (763)

289.—IOA Jordan & Brayton. (253)

883. Ioa vitrea Cope. Vsc. (764)
884. Ioa vigilis Hay. Vs. (764 b.)

290.—BOLEOSOMA De Kay. (254, 255)

885. Boleosoma olmstedi Storer. Vuc. (765)
885 b. *Boleosoma olmstedi atromaculatum* Girard. (Ve.)
885 c. *Boleosoma olmstedi effulgens* Girard. (Vsc.) (767)
885 d. *Boleosoma olmstedi maculatum*[2] Agassiz. Vw. (766)
885 e. *Boleosoma olmstedi ozarcanum*[3] Jordan & Gilbert. Vsw.
885 f. *Boleosoma olmstedi mesæum* Cope. Vw.
885 g. *Boleosoma olmstedi æsopus* Cope. Ve. (760)
886. Boleosoma vexillare Jordan. Ve. (768)
887. Boleosoma susanæ[4] Jordan & Swain. Vs.
888. Boleosoma camurum Forbes. Vw. (770, 771)

291.—ULOCENTRA[5] Jordan. (256)

889. Ulocentra phlox Cope. Vsw. (772)
890. Ulocentra stigmæa Jordan. Vs. (773)
891. Ulocentra simotera Cope. Vs. (774, 775)
892. Ulocentra histrio[6] Jordan & Gilbert. Vsw.
893. Ulocentra blennius[7] Gilbert & Swain. Vs.

292.—DIPLESION Rafinesque. (257)

894. Diplesion blennioides Rafinesque. Vw. (776)

293.—COTTOGASTER Putnam. (258)

895. Cottogaster copelandi Jordan. Vw. (777)
896. Cottogaster putnami Jordan & Gilbert. Vw. (778)

---

[1] CRYSTALLARIA Jordan & Gilbert.

(Genus nova; type *Pleurolepis asprellus* Jordan.)

This genus differs from *Ammocrypta* chiefly in having the premaxillaries non-protractile. The vertical fins are much more developed than in the latter genus, there being 14 dorsal spines, and 12 soft rays in the anal fin. The squamation is much more complete than in *Ammocrypta*, but the body is similarly hyaline. ($K\rho\upsilon\sigma\tau\alpha\lambda\lambda o\varsigma$, crystal.)

[2] I adopt the name *maculatum* for this species or subspecies, the identification of Rafinesque's *Etheostoma nigra* with it being very doubtful. *Pœcilichthys beani* Jordan, Proc. U. S. Nat. Mus., 1884, is identical with *B. maculatum*.

[3] *Boleosoma olmstedi ozarcanum* Jordan & Gilbert, Proc. U. S. Nat. Mus., 1885. Ozark region.

[4] *Boleosoma susanæ* Jordan & Swain, Proc. U. S. Nat. Mus., 1883, 248. Cumberland R., Kentucky.

[5] *Ulocentra atripinnis* Jordan is the adult of *Diplesion simoterum*.

[6] *Etheostoma histrio* Jordan & Gilbert, Proc. U. S. Nat. Mus., 1885. Streams of Arkansas.

[7] *Etheostoma blennius* Gilbert & Swain, Proc. U. S. Nat. Mus., 1884. Streams of Northern Alabama.

CATALOGUE OF THE FISHES OF NORTH AMERICA.

897. **Cottogaster uranidea**[1] Jordan & Gilbert. Vw.
898. **Cottogaster shumardi** Girard. Vsw. (770)

294.—PERCINA Haldeman. (260)

899. **Percina caprodes** Rafinesque. V. (780)
899 b. *Percina caprodes zebra*[2] Agassiz. Vn.

295.—HADROPTERUS Agassiz. (261, 262)

§ *Alvordius* Girard.

900. **Hadropterus macrocephalus** Cope. Vnc. (781)
901. **Hadropterus phoxocephalus** Nelson. Vw. (782)
902. **Hadropterus aspro** Cope & Jordan. Vw. (783)
903. **Hadropterus ouachitæ**[3] Jordan & Gilbert. Vsw.
904. **Hadropterus peltatus**[4] Stauffer. Ve. (784, 785, 786)

§ *Ericosma* Jordan.

905. **Hadropterus evides** Jordan & Copeland. Vw. (787)
906. **Hadropterus fasciatus** Girard. Vsw. (788)

§ *Hadropterus*.

907. **Hadropterus nigrofasciatus** Agassiz. Vs. (790)
908. **Hadropterus aurantiacus** Cope. Vs. (789)
909. **Hadropterus squamatus**[5] Gilbert & Swain. Vs.
910. **Hadropterus cymatotænia**[6] Gilbert & Meek. Vw.
911. **Hadropterus nianguæ**[7] Gilbert & Meek. Vw.
912. **Hadropterus variatus** Kirtland. Vw. (801)

§ *Serraria* Gilbert.

913. **Hadropterus scierus**[8] Swain. Vsw.

§ ——— ?

914. **Hadropterus ? tessellatus** Storer. Vs. (796)
915. **Hadropterus ? cinereus** Storer. Vs. (797)

---

[1] *Cottogaster uranidea* Jordan & Gilbert, Proc. U. S. Nat. Mus., 1885. Washita River, Arkansas.
[2] *Pileoma zebra* Agassiz, Lake Superior, = *Percina manitou* Jordan.
[3] *Hadropterus ouachitæ* Jordan & Gilbert, Proc. U. S. Nat. Mus., 1885. Saline River, Arkansas.
[4] *Hadropterus maculatus* Girard = *Etheostoma peltatum* Stauffer = *Etheostoma nevisense* Cope = *Alvordius crassus* Jordan & Brayton = *Alvordius variatus* Auct. (not *Alvordius maculatus* Girard, nor *Etheostoma variatum* Kirtland).
[5] *Hadropterus squamatus* Gilbert & Swain, Proc. U. S. Nat. Mus., 1885. Tennessee Basin.
[6] *Hadropterus cymatotænia* Gilbert & Meek, Proc. U. S. Nat. Mus., 1885. Ozark region of Missouri.
[7] *Hadropterus nianguæ* Gilbert & Meek Proc. U. S. Nat. Mus., 1885. Niangua River, Southern Missouri.
[8] *Hadropterus scierus* Swain. Proc. U. S. Nat. Mus., 1883, 352. Southern Indiana and southwestward; very abundant in streams of Arkansas and Texas. This species is made the type of a genus, *Serraria*, by Gilbert (Proc. U. S. Nat. Mus., 1884), distinguished from *Hadropterus* by the serrulate preopercle.

296.—**ETHEOSTOMA** Rafinesque. (263, 264, 265, 266)

§ *Rhothæca*[1] Jordan.

916. Etheostoma zonale Cope. Vw. (798)
916 b. *Etheostoma zonale arcansanum*[2] Jordan & Gilbert. Vsw.
917. Etheostoma lynceum[3] Hay. Vs. (799)
918. Etheostoma thalassinum Jordan & Brayton. Vse. (800)
919. Etheostoma inscriptum Jordan & Brayton. Vse. (802)

§ *Nothonotus* Agassiz. (263)

920. Etheostoma camurum[4] Cope. Vc. (791, 795)
921. Etheostoma maculatum[5] Kirtland. Vc. (792, 793)
922. Etheostoma rufolineatum Cope. Vs. (794)

§ *Etheostoma*.

923. Etheostoma flabellare Rafinesque. V. (804)
923 b. *Etheostoma flabellare*[6] *cumberlandicum* Jordan & Swain. Vs.
923 c. *Etheostoma flabellare lineolatum* Agassiz. Vuw. (803)
924. Etheostoma artesiæ Hay. Vs. (809)
925. Etheostoma squamiceps Jordan. S. (805)

§ *Pœcilichthys* Agassiz.

926. Etheostoma virgatum Jordan. Vc. (806)
927. Etheostoma sagitta[7] Jordan & Swain. Vc.
928. Etheostoma saxatile Hay. Vs. (807)
929. Etheostoma rupestre[8] Gilbert & Swain. Vs.
930. Etheostoma luteovinctum[9] Gilbert & Swain. Vs.
931. Etheostoma parvipinne[10] Gilbert & Swain. Vs.
932. Etheostoma boreale[11] Jordan. Vuc.
933. Etheostoma punctulatum[12] Agassiz. Vw.

---

[1] *Rhothæca* Jordan subgenus nova; type *Pœcilichthys zonalis* Cope; substitute for *Nanostoma* Putnam; preoccupied by *Nannostomus* Günther, a genus of Characinidæ ($\rho o \theta o \varsigma$, a current; $o \iota \chi \varepsilon \omega$, to inhabit.) I here regard *Pœcilichthys*, *Nothonotus*, and *Rhothæca* as subgenera under *Etheostoma*.

[2] *Etheostoma zonale arcansanum* Jordan & Gilbert, Proc. U. S. Nat. Mus., 1885. Arkansas and southward.

[3] *Etheostoma lynceum* Hay, nom. sp. nov. for *Nanostoma elegans* Hay; not *Bolcichthys elegans* Girard.

[4] *Pœcilichthys camurus* Cope = *Pœcilichthys vulneratus* Cope.

[5] *Etheostoma maculatum* Kirtland = *Pœcilichthys sanguifluus* Cope.

[6] *Etheostoma cumberlandicum* Jordan & Swain, Proc. U. S. Nat. Mus., 1883, 251. Cumberland River.

[7] *Pœcilichthys sagitta* Jordan & Swain, Proc. U. S. Nat. Mus., 1883, 250. Cumberland River.

[8] *Etheostoma rupestre* Gilbert & Swain, Proc. U. S. Nat. Mus., 1885. Tennessee Basin.

[9] *Etheostoma luteovinctum* Gilbert & Swain, Proc. U. S. Nat. Mus., 1885. Northern Alabama.

[10] *Etheostoma parvipinne* Gilbert & Swain, Proc. U. S. Nat. Mus., 1885. Northern Alabama.

[11] *Pœcilichthys borealis* Jordan, Proc. U. S. Nat. Mus., 1884. Montreal.

[12] This is not the species described as *Pœcilichthys punctulatus* in the Synopsis. For description, see Gilbert & Meek, Proc. U. S. Nat. Mus., 1885. Osage River.

934. **Etheostoma whipplei**[1] Girard. Vsw. (808)
935. **Etheostoma lepidum** Baird & Girard. Vsw. (810)
936. **Etheostoma coeruleum** Storer. Vc. (811)
936 b. *Etheostoma coeruleum spectabile* Agassiz. Vw. (812)
937. **Etheostoma jessiæ**[2] Jordan & Brayton. Vw. (814)
938. **Etheostoma iowæ** Jordan & Meek. Vnw.

§ ———.

939. **Etheostoma tuscumbia**[3] Gilbert & Swain. Vs.

§ *Boleichthys* Girard.

940. **Etheostoma quiescens**[4] Jordan. Vse.
941. **Etheostoma fusiforme**[5] Girard. V. (815, 816, 817, 818, 819, 822)
941 b. *Etheostoma fusiforme eos* Jordan & Copeland. Vnw. (819)
942. **Etheostoma exile**[6] Girard. Vnw. (820, 821)

297.—**ALVARIUS** Girard. (267)

943. **Alvarius lateralis** Girard. Vsw. (823)
944. **Alvarius proeliaris** Hay. Vs. (824)
945. **Alvarius punctulatus** Putnam. Vn. (825)
946. **Alvarius fonticola**[7] Jordan & Gilbert. Vsw.

298.—**PERCA** Linnæus. (268)

947. **Perca lutea** Rafinesque. Vuc. (826)

299.—**STIZOSTEDION** Rafinesque. (269)

948. **Stizostedion vitreum** Mitchill. V. (827)
949. **Stizostedion canadense** Smith. Vne. (828)
949 b. *Stizostedion canadense griseum* De Kay. Vn.
949 c. *Stizostedion canadense boreum* Girard. Vnw.

Family C.—CENTROPOMIDÆ.[8]

300.—**CENTROPOMUS** Lacépède. (270.)

950. **Centropomus undecimalis** Bloch. W. P. (879)

---

[1] This is *P. punctulatus* of the Synopsis, not of Agassiz. It is readily distinguished from the preceding by its slenderer form, larger scales, and less speckled coloration. In life it is spotted with bright red. See Gilbert, l. c.

[2] *Pœcilichthys jessiæ* Jordan & Brayton=*Pœcilichthys asprigenis* Forbes=*Pœcilichthys swaini* Jordan, Proc. U. S. Nat. Mus., 1884, 479. The lateral line in this species is sometimes complete.

[3] *Etheostoma tuscumbia* Gilbert & Swain, Proc. U. S. Nat. Mus., 1885. Tuscumbia Spring, Alabama.

[4] *Pœcilichthys quiescens* Jordan, Proc. U. S. Nat. Mus., 1884, 478. Suwannee River, Georgia.

[5] *Boleosoma fusiformis* Girard=*Boleosoma barratti* Holbrook=*Hololepis erochrous* Cope =*Boleosoma gracile* Girard=*Pœcilichthys butlerianus* Hay=*Pœcilichthys palustris* Gilbert, Proc. U. S. Nat. Mus., 1884, 209. *Pœcilichthys eos* seems also to represent a slight variety of this widely diffused species.

[6] *Boleichthys warreni* is doubtless identical with *Etheostoma exile*. The types of the former are lost.

[7] *Microperca fonticola* Jordan & Gilbert, Proc. U. S. Nat. Mus., 1885. San Marcos Spring, Texas. *Alvarius* and *Microperca* are probably identical.

[8] The characters of the family of *Centropomidæ* are given in detail by Prof. Gill, Proc. U. S. Nat. Mus., 1882, 484.

951. **Centropomus nigrescens**[1] Günther. P.
952. **Centropomus pedimacula**[2] Poey. P. W.
953. **Centropomus robalito**[3] Jordan & Gilbert. P.

Family Cl.—SERRANIDÆ. (86)

301.—ROCCUS Mitchill. (271)

§ *Roccus.*

954. **Roccus septentrionalis**[4] Bloch & Schneider. N. S. Ana. (830)
955. **Roccus chrysops** Rafinesque. Vw. (831)

§ *Morone* (Mitchell) Gill.

956. **Roccus interruptus** Gill. Vsw. (832)
957. **Roccus americanus** Gmelin. N. Ana. (833)

302.—SERRANUS Cuvier. (274)

§ *Centropristis* Cuvier.

958. **Serranus atrarius** Linnæus. S. (836)
959. **Serranus furvus** Walbaum.[5] N. (836 b.)
960. **Serranus philadelphicus**[6] Linnæus. S. (837)

§ *Diplectrum* Holbrook.

961. **Serranus formosus** Linnæus. S. W. (838)
962. **Serranus radialis**[7] Quoy & Gaimard. P. W.

§ *Prionodes* Jenyns.

963. **Serranus subligarius** Cope. W. (839)
964. **Serranus phœbe**[8] Poey. W.

---

[1] *Centropomus nigrescens* Günther, Proc. Zoöl. Soc. London, 1864, 144; Günther, Fishes Centr. Amer., 1869, 407. Mazatlan to Panama.

[2] *Centropomus pedimacula* Poey, Memorias Cuba, II, 1860, 122=*Centropomus medius* Günther, Fish. Centr. Amer., 1869, 406. Both coasts of tropical America, north to Mazatlan.

[3] *Centropomus robalito* Jordan & Gilbert, Proc. U. S. Nat. Mus., 1881, 462. Mazatlan.

[4] This species should stand as above, instead of *Roccus lineatus.* The original *Sciæna lineata* of Bloch was probably one of the European species. To the synonymy add *Perca saxatilis* and *Perca septentrionalis* Bloch & Schneider, Syst. Nat., 1801, 89, 90. *Perca saxatilis* is preoccupied.

[5] *Perca furva* Walbaum, Artedi Piscium, 1279=*Coryphæna nigrescens* Bloch & Schneider, 1801.

[6] *Perca philadelphica* Linnæus, Syst. Nat. X, 291, 1758=ed. XII, 1766, 484=*Perca trifurca* Linnæus, Syst. Nat., ed. XII, 489, 1766.

[7] *Serranus radialis* Quoy & Gaimard, Voyage Freycinet, 316=*Centropristis radialis* Günther, I, 83=*Centropristis macropoma* Günther, Fish. Centr. Amer., 1869, 409. Coast of Brazil and west coast of tropical America, north to Gulf of California.

[8] *Serranus phœbe* Poey.

Light brownish, paler below; a sharply defined white bar extending upward from before vent about to middle of side, its width rather more than diameter of pupil; before this a broad dusky shade extending downward from back; a vaguely defined quadrate paler area below middle of dorsal and another on back of tail; head and fins without sharp markings. Body oblong, the back little elevated, the head large and not sharp

965. **Serranus calopteryx**[1] Jordan & Gilbert. P.
§ *Paralabrax* Girard.
966. **Serranus clathratus** Girard. C. (840)
967. **Serranus maculofasciatus** Steindachner. C. P. (841)
968. **Serranus nebulifer** Girard. C. (842)

303.—HYPOPLECTRUS Gill. (274 b.)

969. **Hypoplectrus nigricans** Poey. W. (843)
970. **Hypoplectrus gemma**[2] Goode & Bean. W.

304.—ANTHIAS[3] Bloch.

971. **Anthias multifasciatus** Gill. P.
972. **Anthias vivanus**[4] Jordan. W.

305.—PARANTHIAS Guichenot. (273 b.)

973. **Paranthias furcifer** Cuv. & Val. W. P. (835 b.)

306.—POLYPRION Cuvier.

974. **Polyprion americanus**[5] Bloch & Schneider. Acc. B. Eu. (835)

307.—STEREOLEPIS Ayres.

975. **Stereolepis gigas** Ayres. C. (834)

---

in profile, much less slender than in *S. subligarius*. Teeth moderate, those on sides of lower jaw and front of upper largest; mouth moderate, the maxillary reaching to center of pupil, 2¼ in head; lower jaw projecting; snout 3⅞ in head; eye large, 3⅞ in head. Scales on cheeks large; preopercle moderately serrate, the teeth nearly uniform; gill-rakers rather short. Caudal moderately forked; dorsal spines rather strong, higher than the soft rays, the longest 2½ in head; second and third anal spines subequal; pectorals reaching front of anal, 1¾ in head; head 2¾; depth 3½; D X, 12, A. III, 7. Scales 5–48–14. L. 8 inches. West Indies, north to Pensacola, Florida. (Poey, Memorias Cuba, I, 1851, 55; *Centropristis phœbe* Günther, I, 85, 1859; *Haliperca phœbe* Poey, Enum. Pisc. Cubens., 1875, 22.)

[1] *Prionodes fasciatus* Jenyns, Voyage of the Beagle, Fishes, 1842, 46 = *Serranus calopteryx* Jordan & Gilbert, Proc. U. S. Nat. Mus., 1881, 350. Mazatlan to Galapagos Islands. The name *fasciatus* is preoccupied in this genus.

[2] *Hypoplectrus gemma* Goode & Bean, Proc. U. S. Nat. Mus., 1882, 428. Garden Key, Florida.

[3] ANTHIAS Bloch.

(*Pronotogrammus* Gill.)

(Bloch, Ichthyologia, type *Labrus anthias* L. = *Anthias sacer* Bloch.)
This genus is closely allied to *Serranus*, differing technically chiefly in the direction of the lateral line, which runs very high and is concurrent with the back, becoming abruptly straight and horizontal below last rays of dorsal. The body is rather strongly compressed, the snout blunt, the mouth oblique, the maxillary broad and scaly, and some of the fins with produced or filamentous rays, and the caudal generally deeply forked. Species of rather small size, mostly inhabiting deep waters.

Anthias multifasciatus = *Pronotogrammus multifasciatus* Gill, Proc. Ac. Nat. Sci. Phila., 1883, 81. Cape San Lucas. See Jordan & Gilbert, Proc. U. S. Nat. Mus., 1882, 360.

[4] *Anthias vivanus* Jordan, Proc. U. S. Nat. Mus., 1885. Pensacola.

[5] *Amphiprion americanus* Bloch & Schneider, Syst. Ichth., 1801, 25; not *Epinephelus oxygeneios* Bloch & Schneider, l. c. 301.

308.—PROMICROPS[1] Gill. (277)

976. Promicrops itaiara Lichtenstein. W. P. (853)

309.—MYCTEROPERCA[2] Gill. (275)

977. Mycteroperca rosacea[3] Streets. P.
978. Mycteroperca falcata phenax[4] Jordan & Swain. W.
979. Mycteroperca microlepis Goode & Bean. W. S. (846)
980. Mycteroperca bonaci[5] Poey. W.
980 b. *Mycteroperca bonaci xanthosticta* Jordan & Swain.
981. Mycteroperca venenosa[6] Linnæus. W. (846 b.)

310.—EPINEPHELUS Bloch. (276)

982. Epinephelus nigritus Holbrook. S. (850)
983. Epinephelus morio Cuv. & Val. S. W. (849)
984. Epinephelus striatus Bloch. W. (850 b.)
985. Epinephelus sellicauda[7] Gill. P.
986. Epinephelus niveatus Cuv. & Val. W. Acc. (851)
987. Epinephelus drummond-hayi Goode & Bean. S. W. (848)
988. Epinephelus apua[8] Bloch. W. (850 c.)
989. Epinephelus ascensionis[9] Osbeck. W. (847)
990. Epinephelus analogus[10] Gill. P.

311.—ALPHESTES[11] Bloch & Schneider.

991. Alphestes multiguttatus Günther. P.

---

[1] *Serranus itaiara* Lichtenstein = *Promicrops guasa* Poey.

For an account of the American genera and species of *Epinephelus* and related forms see Jordan & Swain, Proc. U. S. Nat. Mus., 1884, 358. This paper should supersede the very incomplete account given in the Synopsis.

[2] *Mycteroperca* Gill, 1863 = *Trisotropis* Gill, 1865.

[3] *Epinephelus rosaceus* Streets, Bull. U. S. Nat. Mus., VII, 1877, 51; *M. rosacea* Jordan & Swain, l. c., 361. Gulf of California.

[4] *Mycteroperca falcata phenax* Jordan & Swain, l. c. 363. Key West to Pensacola.

[5] *Serranus bonaci, brunneus, arara*, etc., Poey. See Jordan & Swain. l. c. 370. Key West, southward; Var. *xanthosticta* (l. c. 371) at Pensacola.

[6] *Perca venenosa* L. = *Serranus petrosus* Poey.

[7] *Epinephelus sellicauda* Gill, Proc. Ac. Nat. Sci. Phila., 1862, 250 ; Jordan & Swain, Proc. U. S. Nat. Mus., 1884, 385.

[8] Described in the Synopsis, page 919, under the erroneous name of *Epinephelus guttatus*. See Jordan & Swain, l. c. 389.

[9] Described in the Synopsis, page 539, under the name of *Epinephelus capreolus*. See Jordan & Swain, l. c. 391.

[10] *Epinephelus analogus* Gill, Proc. Ac. Nat. Sci. Phila., 1863. Jordan & Swain, l. c. 393.

[11] ALPHESTES Bloch & Schneider.

(*Prospinus* Poey.)

(Bloch & Schneider, Syst. Ichth., 1801, 236; type, *Epinephelus afer* Bloch.)

This genus includes small species, differing from *Epinephelus* chiefly in the presence of a strong antrorse spine on the lower side of the angle of the preopercle. The three known species are American. (Αλφηστης, enterprising or greedy ; a name applied to some kind of fish which goes in pairs.) *Alphestes multiguttatus* = *Plectropoma multiguttatum* Günther, Proc. Zoöl. Soc. London, 1866, 600. See Jordan & Swain, l. c. 395. Mazatlan to Panama.

312.—ENNEACENTRUS[1] Gill. (276 b.)

§ *Petrometopon* Gill.

992. Enneacentrus guttatus [2] coronatus Cuv. & Val. W.

§ *Enneacentrus.*

993. Enneacentrus tæniops Cuv. & Val. W. Acc. (852 b.)
994. Enneacentrus fulvus ruber [3] Bloch. W.

313.—DERMATOLEPIS [4] Gill.

995. Dermatolepis punctatus Gill. P.

Family CII.—RHYPTICIDÆ.[5]

314.—RHYPTICUS Cuvier. (279)

§ *Rhypticus.*

996. Rhypticus saponaceus [6] Bloch. W.
997. Rhypticus xanti [7] Gill. P.

---

[1] For a statement of the reasons why *Enneacentrus* is preferred to *Bodianus* as the name of this group, see Jordan & Swain, l. c. 397.

[2] *Enneacentrus guttatus* L.; var *coronatus* Cuv. & Val. Key West and southward. For a description of this species see Jordan & Swain, l. c. 398.

[3] The Linnæan name, *Labrus fulvus* (Syst. Nat., X, 1758, 287), has priority for this species. The yellow, red, and brown varieties may stand as *fulvus, ruber,* and *punctatus,* respectively. See Jordan & Swain, Proc. U. S. Nat. Mus., 1884, 402.

Epinephelus fulvus punctatus Linnæus. W. (852b)

[4] DERMATOLEPIS Gill.

(*Lioperca* Gill.)

(Gill, Proc. Ac. Nat. Sci. Phila., 1861, 54; type, *Dermatolepis punctatus* Gill.)
Scales all cycloid; canine teeth very small or obsolete; body comparatively deep; head small; soft dorsal, unusually long, of 19 or 20 rays; spines low. Otherwise essentially as in *Epinephelus.* Two species known. ($\Delta \varepsilon \rho \mu \alpha$, skin; $\lambda \varepsilon \pi \iota s$, scale.)

Dermatolepis punctatus Gill, Proc. Ac. Nat. Sci. Phila., 1861, 54. Jordan & Swain, l. c. 407. Cape San Lucas and adjacent rocky islands.

[5] The genus *Rhypticus,* differing from all other *Serranidæ* in the absence of anal spines and in the reduced number (2 to 4) of the dorsal spines, may be regarded as the type of a distinct family.

[6] *Rhypticus saponaceus* Bloch & Schneider.

Soap-fish; Jabon; Jaboncillo. Olivaceous brown, without distinct markings, in spirits. Body oblong, the back little arched, the snout rather pointed in profile, mouth moderate, the maxillary extending to beyond the eye, $2\frac{1}{4}$ in head; eye about equal to snout, $3\frac{3}{4}$ in head. Opercle with three strong spines, the middle one largest; preopercle with two spines. Head $3\frac{1}{4}$; depth $3\frac{1}{4}$. D. III, 25; A. 17. West Indies, north to Pensacola, Florida.

(*Anthias saponaceus* Bloch & Schneider, Systema Ichth., 1801, 310; Cuv. & Val., III, 63; Günther, I, 172; *Eleutheractis coriaceus* Cope, Trans. Am. Phil. Soc., 1871, 467.)

[7] *Rhypticus xanti* Gill, Proc. Ac. Nat. Sci. Phila., 1862, 250. Cape San Lucas, and southward.

§ *Promicropterus* Gill.

998. Rhypticus bistrispinus[1] Mitchill. S. (855, 857 ?)
999. Rhypticus nigripinnis[2] Gill. P. (856)

Family CIII.—PRIACANTHIDÆ. (87)

315.—PRIACANTHUS Cuvier.

1000. Priacanthus catalufa[3] Poey. W.

316.—PSEUDOPRIACANTHUS[4] Bleeker.

1001. Pseudopriacanthus altus Gill. B. (859)

Family CIV.—LOBOTIDÆ.[5]

317.—LOBOTES Cuvier. (285)

1002. Lobotes surinamensis Bloch. N. S. W. P. (876)

Family CV.—SPARIDÆ.

318.—XENICHTHYS Gill.

1003. Xenichthys xanti[6] Gill. P.

319.—XENISTIUS Jordan & Gilbert. (281)

1004. Xenistius californiensis Steindachner. C. (860)

320.—HOPLOPAGRUS[7] Gill.

1005. Hoplopagrus güntheri Gill. P.

---

[1] *Bodianus bistrispinus* Mitchill, Amer. Monthly Magazine, IV, 1818, 247 (Straits of Bahama)=*Rhypticus maculatus* Holbrook=? *Rhypticus pituitosus* Goode & Bean (young). The specimen from Newport, R. I., recorded by Cope as *Promicropterus decoratus* seems to belong to this species.

[2] *Rhypticus nigripinnis* Gill, 1861. *Rhypticus maculatus* Gill, 1862 = *Promicropterus decoratus* Gill, 1863. Cape San Lucas to Panama.

[3] The species called in the Synopsis *Priacanthus macrophthalmus* (p. 544) and *Priacanthus arenatus* (p. 971) should stand as *Priacanthus catalufa* Poey; Catalufa, Big-eye, Bull's-eye. Instead of the synonymy in the Synopsis, read—
(*Catalufa* Parra, Descr. Dif. Piezas Hist. Nat., 1787 ; *Priacanthus macrophthalmus* Cuv. & Val., III, 95 in part; not *Anthias macrophthalmus* Bloch, which is an East Indian species; *Priacanthus macrophthalmus* Günther, I, 215; *Priacanthus catalufa* Poey, Proc. Ac. Nat. Sci. Phila., 1863, 182; not *Priacanthus arenatus* C. & V.)

[4] *Pseudopriacanthus* Bleeker should be recognized as a genus distinct from *Priacanthus*.

[5] The genus *Lobotes* should be removed from the family of *Sparidæ* and placed in or near the *Serranidæ*, with which it agrees in many respects, differing in the absence of teeth on the vomer. It may stand as a separate family LOBOTIDÆ, which has been defined by Professor Gill, Proc. U. S. Nat. Mus., 1882, 560.

[6] *Xenichthys xanti* Gill, Proc. Ac. Nat. Sci. Phila., 1863, 83 = *Xenichthys xenops* Jordan & Gilbert, Bull. U. S. Fish Com., 1882, 325. Cape San Lucas to Panama.

[7] HOPLOPAGRUS Gill.
(Gill, Proc. Ac. Nat. Sci. Phila., 1862, 253; type *Hoplopagrus güntheri* Gill.)
This genus resembles *Lutjanus* in most respects, differing strikingly in the structure of the anterior nostril and in the dentition. The anterior nostril is remote from the

321.—LUTJANUS[1] Bloch.

1006. **Lutjanus argentiventris**[2] Peters.   P.
1007. **Lutjanus caxis**[3] Bloch & Schneider.   W.
1008. **Lutjanus jocù**[4] Bloch & Schneider.   W.
1009. **Lutjanus griseus**[5] Linnæus.   S. W.   862, 862 b., 864)
1010. **Lutjanus novemfasciatus**[6] Gill.   P.
1011. **Lutjanus guttatus**[7] Steindachner.   P.
1012. **Lutjanus synagris** Linnæus.   W.   (864 b.)
1013. **Lutjanus vivanus**[8] Cuv. & Val.   S. W.   (862 c., 863)
1014. **Lutjanus analis**[9] Cuv. & Val.   W.
1015. **Lutjanus colorado**[10] Jordan & Gilbert.   P.
1016. **Lutjanus aratus**[11] Günther.   P.
1017. **Lutjanus inermis**[12] Peters.   P.

322.—OCYURUS Gill.

1018. **Ocyurus chrysurus**[13] Bloch.   W.   (861)

---

posterior and is placed near the end of the snout; vomer with three large molar teeth; teeth in jaws coarse and blunt. Otherwise as in *Lutjanus*. One species known. ('Οπλος, armed; πάγρος, *Pagrus*, Spanish "Pargo," English "Porgee," a general name for sparoid fishes.)

Hoplopagrus güntheri Gill, l. c. 253; Steindachner, Ichth. Beitr., VI, 1, 1878; Jordan & Swain, Proc. U. S. Nat. Mus., 1884, 429.   Cape San Lucas to Panama.

[1] For a full account of the American species of *Lutjanus* and related genera (*Hoplopagrus, Ocyurus, Rhomboplites, Tropidinius, Aprion, Etelis,* and *Verilus*), see Jordan & Swain, Proc. U. S. Nat. Mus., 1884, 427. The characters of the genera are given by Gill, Proc. U. S. Nat. Mus., 1884, 351, and in the paper above quoted.

[2] *Mesoprion argentiventris* Peters, Berliner Monatsberichte, 1869, 704 = *Lutjanus argentiventris* Jordan & Swain, l. c. 434.   Mazatlan to Panama.

[3] For synonymy and description of *Lutjanus caxis*, see Jordan & Swain, l. c. 435. West Indies, north to Key West.

[4] For synonymy and description of *Lutjanus jocu*, see Jordan & Swain, l. c., 437.

[5] *Labrus griseus* L. = *Anthias caballerote* Bloch & Schneider = *Lutjanus stearnsi* Goode & Bean = *Lutjanus caxis* Synopsis, p. 548; not *Sparus caxis* Bloch & Schneider. The common Gray or Mangrove Snapper of our southern coasts. See Jordan & Swain, l. c. 439.

[6] For synonymy of *Lutjanus novemfasciatus* see Jordan & Swain, l. c. 443. For description see Jordan & Gilbert, Proc. U. S. Nat. Mus., 1881, 232 (*Lutjanus prieto* J. & G.). Cape San Lucas to Panama.

[7] For synonymy and description of *Lutjanus guttatus*, see Jordan & Swain, l. c. 447. Mazatlan to Panama.

[8] *Mesoprion vivanus* Cuv. & Val.= *Mesoprion campechanus* Poey= *Lutjanus blackfordi* Goode & Bean.   Charleston and Pensacola to Aspinwall and the Lesser Antilles. For synonymy and description of *Lutjanus vivanus*, see Jordan & Swain, l. c. 453.

[9] For synonymy and description of *Lutjanus analis*, see Jordan & Swain, l. c. 455. West Indies, north to Key West.

[10] For synonymy and description of *Lutjanus colorado*, see Jordan & Gilbert, Proc. U. S. Nat. Mus. 1881, 338, and Jordan & Swain, l. c. 1884, 457.   Mazatlan to Panama.

[11] For synonymy and description of *Lutjanus aratus*, see Jordan & Swain, l. c. 460. Mazatlan to Panama.

[12] For synonymy and description of *Lutjanus inermis*, see Jordan & Swain, l. c. 459. One specimen known, from Mazatlan.

[13] For synonymy and detailed description of *Ocyurus chrysurus*, see Jordan & Swain, Proc. U. S. Nat. Mus., 1884, 461.

323.—RHOMBOPLITES Gill.

1019. Rhomboplites aurorubens [1] Cuv. & Val. W. S. (865)

324.—CONODON Cuv. & Val. (282 b.)

1020. Conodon nobilis Linnæus. W. (866)
1021. Conodon serrifer [2] Jordan & Gilbert. P.

325.—ORTHOPRISTIS [3] Girard.

§ *Microlepidotus* Gill.

1022. Orthopristis inornatus [4] Gill. P.

§ *Orthopristis.*

1023. Orthopristis brevipinnis [5] Steindachner. P.
1024. Orthopristis cantharinus [6] Jenyns. P.
1025. Orthopristis chalceus [7] Günther. P.
1026. Orthopristis chrysopterus [8] Linnæus. S. W. (867, 868)

326.—POMADASYS Lacépède. (283)

§ *Hæmulopsis* Steindachner.

1027. Pomadasys leuciscus [9] Günther. P.
1028. Pomadasys elongatus [10] Steindachner. P.
1029. Pomadasys nitidus [11] Steindachner. P.
1030. Pomadasys axillaris [12] Steindachner. P.

---

[1] For synonymy and description of *Rhomboplites aurorubens*, see Jordan & Swain, l. c. 464.

[2] *Conodon serrifer* Jordan & Gilbert, Proc. U. S. Nat. Mus., 1882, 351. Boca Soledad, Lower California.

[3] It is probably better to regard *Conodon, Orthopristis,* and *Anisotremus* as generically distinct from *Pomadasys.* See Jordan & Gilbert, Proc. U. S. Nat. Mus., 1881, 384, for an analysis of the characters of the Pacific coast species of this group.

[4] *Microlepidotus inornatus* Gill, Proc. Ac. Nat. Sci. Phila., 1862, 256. Cape San Lucas (not *Pomadasys inornatus* Jordan & Gilbert, l. c. 388).

[5] *Pristipoma brevipinne* Steindachner, Ichthyol. Notizen, VIII, 1869, 10. Mazatlan to Panama. See Jordan & Gilbert, Proc. U. S. Nat. Mus., 1882, 625.

[6] *Pristipoma cantharinum* Jenyns, Zoöl. Voy. Beagle, 49, 1842, and Günther, 1, 363, Günther's description agrees with a specimen from Guaymas, diagnosed by Jordan & Gilbert, Proc. U. S. Nat. Mus., 1881, 274 as "*Pomadasys? inornatus,*" and on page 388, l. c., as *P. cantharinus.* This species is distinct from *O. chalceus,* and is probably the original *cantharinus* from the Galapagos Islands. I have, however, seen specimens of *O. chalceus* from the Galapagos.

[7] For synonymy and diagnosis of *Orthopristis chalceus* see Jordan & Gilbert, Proc. U. S. Nat. Mus., 1881, 387. Mazatlan to Galapagos Islands.

[8] *Perca chrysoptera* Linn. Syst. Nat.=*Pristipoma fulvomaculatum* and *P. fasciatum* of Cuv. & Val. The Linnæan type, sent by Dr. Garden from Charleston, has been identified by Dr. Bean.

[9] For diagnosis see Jordan & Gilbert, l. c. 387. Mazatlan to Panama.

[10] *Pristipoma leuciscus* var. *elongatus,* Steindachner, Neue & Seltene Fische aus K. K. Museum, Wien, &c., 1879, taf. 9, f. 2. *Pomadasys elongatus* Jordan & Gilbert, Proc. U. S. Nat. Mus., 1882, 352. Mazatlan to Panama.

[11] For diagnosis of *Pomadasys nitidus* see Jordan & Gilbert, l. c. 387. Mazatlan to Panama.

[12] For diagnosis of *Pomadasys axillaris* see Jordan & Gilbert, l. c. 387. Gulf of California to Panama.

§ *Pseudopristipoma* Sauvage.

1031. **Pomadasys panamensis**[1] Steindachner. P.

§ *Pomadasys.*

1032. **Pomadasys branicki**[2] Steindachner. P.
1033. **Pomadasys macracanthus**[3] Günther. P.

327.—ANISOTREMUS Gill.

1034. **Anisotremus dovii**[4] Günther. P.
1035. **Anisotremus cæsius**[5] Jordan & Gilbert. P.
1036. **Anisotremus interruptus**[6] Gill. P. (871 b.)
1037. **Anisotremus bilineatus** Cuv. & Val. W. (871)
1038. **Anisotremus davidsoni** Steindachner C. (869)
1039. **Anisotremus virginicus** Linnæus. W. (870)
1039 b. *Anisotremus virginicus*[7] *tæniatus* Gill. P.

328.—HÆMULON[8] Cuvier.

§ *Orthostœchus* Gill.

1040. **Hæmulon maculicauda**[9] Gill. P.

§ *Lythrulon* Jordan & Swain.

1041. **Hæmulon flaviguttatum**[10] Gill. P.

§ *Bathystoma* Scudder.

1042. **Hæmulon aurolineatum**[11] Cuv. & Val. W. (874 b.)
1043. **Hæmulon rimator**[12] Jordan & Swain. S. W. (873)

---

[1] For diagnosis of *Pomadasys panamensis* see Jordan and Gilbert, l. c. 387. Mazatlan to Panama.

[2] For diagnosis of *Pomadasys branicki* see Jordan and Gilbert, l. c. 386. Mazatlan to Tumbez, Peru.

[3] For diagnosis of *Pomadasys macracanthus* see Jordan & Gilbert, l. c. 386. Mazatlan to Panama.

[4] For diagnosis of *Anisotremus dovii* see Jordan & Gilbert, l. c. 386. Mazatlan to Panama.

[5] *Pomadasys cæsius* Jordan & Gilbert, Proc. U. S. Nat. Mus., 1881, 383. Mazatlan.

[6] *Anisotremus modestus* Tschudi, accredited to Mazatlan (as *Pristipoma notatum*), by Peters, is here omitted, for reasons given in Proc. Ac. Nat. Sci. Phila., 1883, 286.

[7] *Anisotremus tæniatus* Gill. Proc. Ac. Nat. Sci. Phila., 1861, 107. Gulf of California to Panama. For characters of this subspecies see Jordan & Gilbert, Proc. U. S. Nat. Mus., 1882, 372.

[8] The generic name *Diabasis* is preoccupied and must give place to *Hæmulon*. For a detailed account of the species of this genus see Jordan & Swain, Proc. U. S. Nat. Mus., 1884, 281.

[9] For an account of *Hæmulon maculicauda* see Jordan & Swain, l. c. 315. Cape San Lucas to Panama.

[10] See Jordan & Swain, l. c. 314. Cape San Lucas to Panama.

[11] *Hæmulon aurolineatum* Cuv. & Val. = *Hæmulon jeniguano* Poey. See Jordan & Swain, l. c. 310.

[12] *Hæmulon rimator* Jordan & Swain, l. c., 308. = *Hæmulon chrysopterum* C. & V., not of L.

§ *Brachygenys* Scudder.

1044. Hæmulon tæniatum[1] Poey. W.

§ *Hæmulon.*

1045. Hæmulon flavolineatum[2] Desmarest. W.
1046. Hæmulon plumieri Lacépède. S. W. (872)
1047. Hæmulon sciurus[3] Shaw. W. (872 b.)
1848. Hæmulon steindachneri[4] Jordan & Gilbert. P.
1049. Hæmulon fremebundum[5] Goode & Bean. W. (874)
1050. Hæmulon scudderi[6] Gill. P.
1051. Hæmulon acutum[7] Poey. W. (873 b.)
1052. Hæmulon gibbosum[8] Walbaum. W. (873 c.)
1053. Hæmulon sexfasciatum[9] Gill. P.

329.—SPARUS Linnæus.

§ *Pagrus* Cuv. & Val.

1054. Sparus pagrus Linnæus. S. Eu. (878)

330.—CALAMUS Swainson. (285)

1055. Calamus proridens[10] Jordan & Gilbert. W. (876 b.)
1056. Calamus calamus[11] Cuv. & Val. W.
1057. Calamus bajonado[12] Bloch & Schneider. W.
1058. Calamus brachysomus[13] Lockington. P.

---

[1] For description of *Hæmulon tæniatum* see Jordan & Swain, l. c. 307. West Indies, north to Key West.

[2] For description and synonymy of *Hæmulon flavolineatum* see Jordan & Swain, l. c. 305. West Indies north to Key West.

[3] *Sparus sciurus* Shaw = *Hæmulon elegans* Cuvier. See Jordan & Swain, l. c. 301.

[4] *Diabasis steindachneri* Jordan & Gilbert, Bull. U. S. Fish Com., 1881, 322. Mazatlan to Panama.

[5] For description of the adult form of *Hæmulon fremebundum* see Jordan & Swain, l. c. 297. This species has been recently described from Jamaica under the name of *Diabasis lateralis* (Vaillant & Bocourt, Miss. Sci. au Mexique, 1883.)

[6] For description of *Hæmulon scudderi* see Jordan & Swain, l. c. 296. Mazatlan to Panama.

[7] Described by Jordan & Swain, l. c. 294.

[8] For description of *Hæmulon gibbosum* see Jordan & Swain, l. c. 290. The oldest binomial name of this species is that of *Perca gibbosa* Walbaum, Artedi, Piscium, 1792, 348, based on *Perca marina gibbosa*, the Margate-fish, of Catesby.

[9] For description of *Hæmulon sexfasciatum* see Jordan & Swain, l. c. 288.

[10] *Calamus proridens* Jordan & Gilbert, Proc. U. S. Nat. Mus., 1884, 239 = *Calamus pennatula* Jordan & Gilbert, Proc. U. S. Nat. Mus., 1884, 15 (not of Guichenot). West Indies, north to Key West. For synonymy and description of this and other species of *Calamus* see Jordan & Gilbert, Proc. U. S. Nat. Mus., 1884, 15.

[11] For synonymy and description of *Calamus calamus* see Jordan & Gilbert, l. c. 16. West Indies, north to Key West.

[12] For synonymy and description of *Calamus bajonado* see Jordan & Gilbert, l. c. 20. West Indies, north to Key West.

[13] *Sparus brachysomus* Lockington, Proc. U. S. Nat. Mus., 1880, 284. Magdalena Bay, southward.

1059. **Calamus leucosteus**[1] Jordan & Gilbert. S. (876 c.)
1060. **Calamus penna**[2] Cuv. & Val. S. W. (877)
1061. **Calamus arctifrons** Goode & Bean. S. W. (876 e.)

331.—STENOTOMUS Gill.

1062. **Stenotomus caprinus** Bean. S. (881 b.)
1063. **Stenotomus chrysops**[3] Linnæus. N. S. (881)
1063 b. *Stenotomus chrysops aculeatus* Cuv. & Val. N. S. (880)

332.—DIPLODUS Rafinesque. (267)

§ *Lagodon* Holbrook.

1064. **Diplodus rhomboides** Linnæus. S. W. (882)
1065. **Diplodus unimaculatus**[4] Bloch. W. (885 b.)

§ *Archosargus* Gill.

1066. **Diplodus probatocephalus** Walbaum. N. S. (883)

§ *Diplodus*.

1067. **Diplodus holbrooki** Bean. S. (884, 885)

333.—GIRELLA Gray. (268)

1068. **Girella nigricans** Ayres. C. (886)

---

[1] *Calamus leucosteus* Jordan & Gilbert nom. sp. nov. "White Bone Porgy." Body formed much as in *Calamus penna*, short and deep, with steep anterior profile and high, arched back, the profile nearly straight from snout to above eyes, thence convex. Head deeper than long; the preorbital region very deep, its least depth $2\frac{1}{4}$ in head, half greater than interorbital width. Eye rather large, $2\frac{2}{3}$ in head in adults; a strong blunt prominence before it. Mouth rather large, the maxillary $2\frac{2}{3}$ in head. Outer teeth in both jaws moderately enlarged, canine-like, about ten in each jaw, none of them directed forwards. Highest dorsal spine $2\frac{1}{8}$ in head. Pectorals very long, $2\frac{3}{4}$ in length of body. Ventrals $1\frac{3}{4}$ in head. Scales large, those on cheeks in five rows. Smutty-silvery sides with vague cross bars; dorsal and anal fins with dark blotches; ventrals dusky; no black axillary spot. Head $2\frac{1}{2}$; depth $2\frac{1}{4}$. D. XII, 12; A. III, 10. Scales 7-51-14. Length about a foot. Charleston, S. C.

[2] *Pagellus penna* Cuv. & Val. = *Pagellus milneri* Goode & Bean. For synonymy and description of *Calamus penna* see Jordan & Gilbert, l. c. 21.

[3] According to Dr. Bean, the types of *Sparus chrysops* and *Sparus argyrops* Linnæus are both the common scup. The large or Southern scup, if really a distinct species or variety, should stand as *Stenotomus aculeatus* Cuv. & Val.

[4] *Diplodus unimaculatus* (Bloch). *Salema; Bream.*
This species has the teeth emarginate, as in *D. rhomboides*, and it likewise belongs to the subgenus *Lagodon*. It is distinguished from *D. rhomboides* by its deeper body, and by the longer second anal spine, which extends beyond the tip of the third spine when depressed. It has, further, 13 dorsal spines instead of 12, and its coloration is deeper and more golden. West Indies, north to Pensacola.
To the synonymy add:
(*Salema* Marcgrave, Hist. Brazil, p. 153; *Perca unimaculata* Bloch, taf. 308; *Sargus unimaculatus* Cuv. & Val., VI, 62, 1830; *Sargus unimaculatus* Günther, I, 446; *Sargus caribæus* Poey, Memorias Cuba, II, 1860, 198; *Diplodus unimaculatus* Jordan, Proc. U. S. Nat. Mus., 1884, 126.)

334.—KYPHOSUS Lacépède. (289)

1069. **Kyphosus sectatrix**[1] Linnæus. W. S. (887)
1070. **Kyphosus analogus**[2] Gill. P.

335.—CÆSIOSOMA[3] Kaup. (290)

1071. **Cæsiosoma californiense** Steindachner. S. (888)

Family CVI.—CIRRHITIDÆ.[4]

336.—CIRRHITES Lacépède.

1072. **Cirrhites rivulatus** Valenciennes. P.

Family CVII.—APOGONIDÆ.

337.—APOGON Lacépède. (291)

§ *Apogon.*

1073. **Apogon imberbis**[5] Linnæus. Eu. N. (Acc.) (889)
1074. **Apogon maculatus** Poey. W. (889 b.)
1075. **Apogon retrosella**[6] Gill. P.

§ *Apogonichthys* Bleeker.

1076. **Apogon alutus** Jordan & Gilbert. W. (889 c.)

§ *Glossamia* Gill.

1077. **Apogon pandionis** Goode & Bean. B. (890)

Family CVIII.—MULLIDÆ.

338.—MULLUS Linnæus. (292)

1078. **Mullus barbatus (L.) auratus** Jordan & Gilbert. S. N. Eu. (891)

---

[1] *Perca sectatrix* L., Syst. Nat., Ed. XII, 486 = *Pimelepterus bosci* Cuv. & Val.

[2] *Pimelepterus analogus* Gill, Proc. Ac. Nat. Sci. Phila., 1862, 245. Mazatlan to Panama.

[3] I now adopt the genus *Cæsiosoma* for *Scorpis californiensis*. This species differs much from the figure of *Scorpis geórgianus*, to which it may not be really related. *Cæsiosoma* is certainly not a *Chætodont*, but a very near relative of *Kyphosus*. The propriety of placing *Girella*, *Kyphosus*, and *Cæsiosoma* among the *Sparidæ* is questionable. Gill has placed them together in his family *Pimelepteridæ*.

[4] See Günther, ii, 70, for the characters of the family of *Cirrhitidæ* and of the genus *Cirrhites*. Our species, *Cirrhites rivulatus* Valenciennes, Voyage Vénus Poiss., 399 = *Cirrhitichthys rivulatus* Günther, Fish. Centr. Amer., 1869, 421 = *Cirrhites betaurus* Gill, Proc. Ac. Nat. Sci. Phila., 1862, is found from Cape San Lucas to the Galapagos Islands.

[5] The specimen from Newport, R. I., recorded by Cope as *Apogon americanus*, belongs to the European species, *Apogon imberbis* L. It has been compared with the latter, at my request, by Mr. S. E. Meek.

[6] *Amia retrosella* Gill, Proc. Ac. Nat. Sci. Phila., 1862, 251. Cape San Lucas.

339.—UPENEUS Cuvier. (293)

1079. Upeneus maculatus Bloch. W. (892)
1080. Upeneus martinicus[1] Cuv. & Val. W.
1081. Upeneus grandisquamis[2] Gill. P.
1082. Upeneus dentatus[3] Gill. P.

Family CIX.—SCIÆNIDÆ. (91)

340.—APLODINOTUS Rafinesque. (294)

1083. Aplodinotus grunniens Rafinesque. V. (893)

341.—POGONIAS Lacépède. (295)

1084. Pogonias chromis Linnæus. S. (894)

342.—RONCADOR Jordan & Gilbert. (296 b.)

1085. Roncador stearnsi Steindachner. C. (899)

343.—SCIÆNA Linnæus. (296)

§ *Stelliferus* Stark.

1086. Sciæna lanceolata Holbrook. S. (895)

§ *Bairdiella* Gill.

1087. Sciæna chrysura Lacépède. S. (896)
1088. Sciæna icistia[4] Jordan & Gilbert. P.

§ *Sciæna*.

1089. Sciæna jacobi Steindachner. C. (897)
1090. Sciæna sciera[5] Jordan & Gilbert. P.
1091. Sciæna ocellata Linnæus. S. (898)

344.—JOHNIUS[6] Bloch. (296 c.)

§ *Corvina* Cuvier.

1092. Johnius saturnus Girard. C. (900)

---

[1] *Upeneus martinicus* Cuv. & Val.
Yellow Goat-fish: *Salmonete amarilla*. Red; sides with a broad longitudinal band of bright yellow; snout with yellow streaks; vertical fins and patches on sides of head bright yellow. Body moderately elongate; anterior profile gibbous before the eyes; eyes large, 3½ in head. Teeth bluntish, rather strong, in two or three series, the lower larger than the upper; no teeth on vomer. Interorbital space flat, 3⅔ in head. Barbels 1¾ in head; longest dorsal spine 1½; anal small. Head 3½; depth 4, D. VII-9; A. 7. Scales 2½-37-7. L. 1 foot. West Indies, north to Key West.
(*Upeneus martinicus* and *U. balteatus* Cuvier & Valenciennes, III, 484, 1829; *Upeneus flavovittatus* Poey, Memorias Cuba, I, 224, 1856; *Mulloides flavovittatus* Günther, I, 403.)

[2] *Upeneus grandisquamis* Gill, Proc. Ac. Nat. Sci. Phila., 1863, 168 = *Upeneus tetraspilus* Günther, Fish. Centr. Amer., 1869, 420. Mazatlan to Panama.

[3] *Upeneus dentatus* Gill, Proc. Ac. Nat. Sci. Phila., 1862, 256; Jordan & Gilbert, Proc. U. S. Nat. Mus., 1882, 363. Cape San Lucas.

[4] *Sciæna icistia* Jordan & Gilbert, Proc. U. S. Nat. Mus., 1881, 356. Mazatlan.

[5] *Sciæna sciera* Jordan & Gilbert, Proc. U. S. Nat. Mus., 1884, 480. Mazatlan to Panama.

[6] The name *Johnius* Bloch & Schneider should be used instead of *Corvina* (pp. 572, 932) for the section of *Sciæna* characterized by the absence of bony serræ on the preopercle. The intergradations among the species will perhaps prevent this group from being considered as a genus from *Sciæna*.
Johnius Bloch & Schneider, Syst. Ichth., 1801, p. 74; type (as restricted by Cuvier & Gill) *Johnius carutta* Bloch. (Named for John, a missionary in Tranquebar.)

345.—EQUES Bloch. (296 d.)

§ *Pareques* Gill.

1093. Eques acuminatus [1] Bloch & Schneider. W. (901 b.)

§ *Eques.*

1094. Eques lanceolatus Gmelin. W. (901 b.)

346.—LIOSTOMUS Lacépède. (297)

1095. Liostomus xanthurus Lacépède. S. (902)

347.—LARIMUS Cuvier & Valenciennes. (302)

1096. Larimus fasciatus Holbrook. S. (911)
1097. Larimus breviceps [2] Cuv. & Val. P. W.

348.—GENYONEMUS Gill. (298)

1098. Genyonemus lineatus Ayres. C. (903)

349.—MICROPOGON Cuv. & Val. (299)

1099. Micropogon undulatus Linnæus. N. S. (904)
1100. Micropogon ectenes [3] Jordan & Gilbert. P.

350.—UMBRINA Cuvier. (300)

1101. Umbrina roncador Jordan & Gilbert. C. (905)
1102. Umbrina xanti [4] Gill. P.
1103. Umbrina dorsalis [5] Gill. P.
1104. Umbrina broussoneti Cuv. & Val. W. (906)

351.—MENTICIRRUS Gill. (301)

1105. Menticirrus littoralis Holbrook. S. (908)
1106. Menticirrus elongatus [6] Günther. P.
1107. Menticirrus undulatus Girard. C. (910)
1108. Menticirrus saxatilis [7] Bloch & Schneider. N. S. (907)
1109. Menticirrus alburnus Linnæus. S. (909)
1110. Menticirrus panamensis [8] Steindachner. P.
1111. Menticirrus nasus [9] Günther. P.

---

[1] The subgenus *Pareques* and its typical species *Sciæna acuminata* should be transferred to the genus *Eques.*

[2] *Larimus breviceps* Cuv. & Val., V, 146; Günther, I, 268. Both coasts of Tropical America, north to Mazatlan.

[3] *Micropogon ectenes* Jordan & Gilbert, Proc. U. S. Nat. Mus., 1881, 355; 1882, 282. Mazatlan.

[4] *Umbrina xanti* Gill, Proc. Ac. Nat. Sci. Phila., 1862, 257 = *Umbrina analis* Günther, Fish. Centr. Amer., 1869, 426. For diagnosis, see Jordan & Gilbert, Proc. U. S. Nat. Mus., 1882, 364.

[5] *Umbrina dorsalis* Gill, l. c. 1862, 257. See Jordan & Gilbert, l. c. 364.

[6] *Umbrina elongata* Günther, Proc. Zoöl. Soc. London, 1864, 148. For diagnosis see Jordan & Gilbert, l. c. 284. Mazatlan to Panama.

[7] The name *Johnius saxatilis* (Bloch & Schneider, Syst. Ichth., 1801, 75, based on a specimen from New York, now in the museum at Berlin) has priority for the species called in the Synopsis, *Menticirrus nebulosus.*

[8] *Umbrina panamensis* Steindachner, Ichth. Beitr., IV, 9, 1875. Mazatlan to Panama. See Jordan & Gilbert, l. c. 284.

[9] *Umbrina nasus* Günther, Fish. Centr. Amer., 1869, 426. Mazatlan to Panama. See Jordan & Gilbert, l. c. 284.

352.—CYNOSCION Gill. (303, 304)

§ *Atractoscion* Gill.

1112. Cynoscion nobile Ayres. C. (912)

§ *Cynoscion.*

1113. Cynoscion regale Bloch & Schneider. N. S. (915)
1114. Cynoscion thalassinum Holbrook. S. (916)
1115. Cynoscion nothum Holbrook. S. (914)
1116. Cynoscion othonopterum[1] Jordan & Gilbert. P.
1117. Cynoscion parvipinne Ayres. C. P. (913)
1118. Cynoscion xanthulum[2] Jordan & Gilbert. P.
1119. Cynoscion reticulatum[3] Günther. P.
1120. Cynoscion maculatum Mitchill. S. (917)

353.—SERIPHUS Ayres. (305)

1121. Seriphus politus Ayres. C. (918)

Family CX.—GERRIDÆ. (92)

354.—GERRES Cuvier. (306)

§ *Gerres.*

1122. Gerres plumieri Cuv. & Val. W. (919)
1123. Gerres lineatus[4] Humboldt. P.
1124. Gerres olisthostoma Goode & Bean. S. W. (919 b.)
1125. Gerres peruvianus[5] Cuv. & Val. P.

§ *Diapterus* Ranzani.

1126. Gerres cinereus Walbaum. PW. (921 b.)
1127. Gerres californiensis Gill. P.
1128. Gerres gula[6] Cuv. & Val. S. W. (920, 921)
1129. Gerres gracilis[7] Gill. P. W. S. (922)
1130. Gerres jonesi Günther. W.
1131. Gerres lefroyi[8] Goode. W.

---

[1] *Cynoscion othonopterum* Jordan & Gilbert, Proc. U. S. Nat. Mus., 1881, 274. Gulf of California.

[2] *Cynoscion xanthulum* Jordan & Gilbert, Proc. U. S. Nat. Mus., 1881, 460. Mazatlan.

[3] *Otolithus reticulatus* Günther, Proc. Zoöl. Soc. London, 1864, 149. Mazatlan to Panama. For diagnosis of this and other species of Cynoscion see Jordan & Gilbert, Bull. U. S. Fish Comm., 1881, 319.

[4] For synonymy and description of *Gerres lineatus*, see Jordan & Gilbert, Proc. U. S. Mus., 1881, 330. Mazatlan to Panama.

[5] For synonymy and diagnosis of *Gerres peruvianus*, see Jordan & Gilbert, Bull. U. S. Fish Comm., 1881, 330. Mazatlan to Peru. For a detailed account of American species of *Gerres*, see Evermann & Meek, Proc. Ac. Nat. Sci. Phila., 1883, 116.

[6] *Gerres homonymus* seems to me indistinguishable from *Gerres gula*.

[7] *Diapterus gracilis* Gill. Proc. Ac. Nat. Sci. Phila., 1882, 246 = *Diapterus harengulus* Goode & Bean. Abundant on both coasts of Tropical America.

To its synonymy add:

(*Diapterus gracilis* Gill, Proc. Ac. Nat. Sci. Phila, 1862, 246; *Eucinostomus pseudogula* Poey, Enum. Pisc. Cubens., 124, 1875; Jordan & Gilbert, Bull. U. S. Fish Comm., 1881, 329; Evermann & Meek, Proc. Ac. Nat. Sci. Phila., 1883, 118. *Gerres aprion* Günther, IV, 255, 1862, not of C. & V.)

[8] *Gerres lefroyi* Goode. Bluish above the back, rather darker than in related species, with oblique dusky cross shades; faint dusky streaks along sides; lower parts

Family CXI.—EMBIOTOCIDÆ. (93)

355.—HYSTEROCARPUS Gibbons. (307)

1132. **Hysterocarpus traski** Gibbons. T. (923)

356.—ABEONA Girard. (308)

1133. **Abeona minima** Gibbons. C. (924)
1134. **Abeona aurora** Jordan & Gilbert. C. (925)

357.—BRACHYISTIUS Gill. (308 b.)

1135. **Brachyistius frenatus** Gill. C. (926)
1136. **Brachyistius rosaceus** Jordan & Gilbert. C. (927)

358.—MICROMETRUS Gibbons. (309)

1137. **Micrometrus aggregatus** Gibbons. C. (928)

359.—HOLCONOTUS Agassiz. (310)

§ *Hypocritichthys* Gill.

1138. **Holconotus analis** Alex. Agassiz. C. (929)

§ *Hyperprosopon* Gibbons.

1139. **Holconotus argenteus** Gibbons. C. (930)
1140. **Holconotus agassizii** Gill. C. (931)

§ *Holconotus.*

1141. **Holconotus rhodoterus** Agassiz. C. (933)

360.—AMPHISTICHUS Agassiz. (310 b.)

1142. **Amphistichus argenteus** Agassiz. C. (933)

361.—HYPSURUS Alex. Agassiz. (311)

1143. **Hypsurus caryi** Agassiz. C. (934)

362.—DITREMA Schlegel. (312)

§ *Tæniotoca* Alex. Agassiz.

1144. **Ditrema laterale** Agassiz. C. (935)

---

brightly silvery; tip of spinous dorsal usually black, other fins pale; slenderer than any other of the American species; the snout rather sharp; the outlines of the body not angular; eye rather large, 3 in head, nearly equal to the flattish interorbital space; premaxillary groove linear, naked, formed as in *G. gracilis*; fins low; the longest dorsal spines, 2 in head; anal spines short; pectoral short, 1¼ in head; head, 3⅛; depth, 3½; D, IX, 10; A, II, 8; scales, 4—45—10; L., 4 inches. West Indies, north to Cedar Key, Florida. Well distinguished from all related species by the presence of but two anal spines. The only other species with two anal spines is *G. rhombeus* C. & V., an ally of *G. olisthostoma*.

(*Diapterus lefroyi* Goode, Am. Journ. Sci. Arts, 1874, 123; *Eucinostomus lefroyi* Goode, Bull. U. S. Nat. Mus. V., 1876, 39; *Eucinostomus productus* Poey, Ann. Lyc. N. Y., XI, 59, 1876; Evermann & Meek, Proc. Ac. Nat. Sci. Phila., 1883, 118.)

[97] CATALOGUE OF THE FISHES OF NORTH AMERICA.

§ *Embiotoca* Agassiz.

1145. **Ditrema jacksoni** Agassiz. C. (936)

§ *Phanerodon* Girard.

1146. **Ditrema atripes** Jordan & Gilbert. C. (937)
1147. **Ditrema furcatum** Girard. C. (938)

363.—RHACOCHILUS Agassiz. (313)

1148. **Rhacochilus toxotes** Agassiz. C. (939)

364.—DAMALICHTHYS Girard. (314)

1149. **Damalichthys argyrosomus** Girard. C. (940)

Family CXII.—LABRIDÆ. (94)

365.—CTENOLABRUS Cuv. & Val. (315)

§ *Tautogolabrus* Günther.

1150. **Ctenolabrus adspersus** Walbaum. N. (941)

366.—HIATULA Lacépède. (316)

1151. **Hiatula onitis** Linnæus. N. (948)

367.—LACHNOLÆMUS Cuv. & Val. (317)

1152. **Lachnolæmus maximus**[1] Walbaum. W. (943)

368.—BODIANUS[2] Bloch. (318)

1153. **Bodianus rufus** Linnæus. W. (944)
1154. **Bodianus diplotænia**[3] Gill. P.
1155. **Bodianus pectoralis**[4] Gill. P.

---

[1] The species commonly known as *Lachnolæmus falcatus* must stand as *Lachnolæmus maximus* Walbaum.

The *Labrus falcatus* of Linnæus is certainly not this species as supposed by Valenciennes, but is probably some species of *Trachynotus*. The oldest name, certainly, belonging to the *Lachnolæmus* is that of *Labrus maximus* Walbaum, Artedi Piscium, 1792, 261 = (*Lachnolæmus suillus* Cuvier, Règne Animal, Ed. II, 1829, 257, both names based on *Suillus*, the hog-fish of Catesby.)

[2] The genus called in the text *Harpe* must probably stand as

BODIANUS Bloch.

(Bloch, Ichthyologia, about 1780; type *Bodianus bodianus* Bloch = *Labrus rufus* L.)
The genus *Bodianus* Bloch is a medley of unrelated fishes. The group was, however, based especially on *Bodianus bodianus* Bloch, from the Portuguese name, of which (*Bodiano* or *Pudiano*) the name *Bodianus* was derived.

[3] *Harpe diplotænia* Gill, Proc. Ac. Nat. Sci. Phila., 1862, 140; Jordan & Gilbert, Proc. U. S. Nat. Mus., 1882, 367. Cape San Lucas.

[4] *Harpe pectoralis* Gill, l. c. 141. Gulf of California southward. This is probably the male of *Bodianus diplotænia*.

369.—DECODON[1] Günther.

1156. Decodon puellaris Poey. W.

370.—TROCHOCOPUS Günther. (318b.)

§ *Pimelometopon* Gill.

1157. Trochocopus pulcher Ayres. C. (945)

371.—PLATYGLOSSUS Bleeker. (319)

1158. Platyglossus radiatus[2] Linnæus. W. (946)
1159. Platyglossus bivittatus[3] Bloch. S. W. (947; 948)
1160. Platyglossus caudalis Poey. W. (948 b.)

---

[1] DECODON Günther.

(Günther, Cat. Fish. Brit. Mus., IV, 101, 1862; type *Cossyphus puellaris* Poey.)
Body moderately compressed, oblong, covered with large scales; head oblong; cheeks, opercles, and lower limb of preopercle scaly, the posterior limb being naked; base of dorsal and anal not scaly; lateral line continuous. Teeth essentially as in *Harpe*, those of the jaws in a single series; four canines in the front of each jaw; a posterior canine on each premaxillary. Dorsal with eleven spines; anal with three. A single species, intermediate between *Bodianus* and *Trochocopus*, having the large scales of the former and the naked fins of the latter. Apparently the genera in this group have been too much subdivided. ($\Delta \varepsilon \varkappa \alpha \varsigma$, ten; $\dot{o}\delta o\dot{v}\varsigma$, tooth; there being ten canines.)

Decodon puellaris.

Rose-colored, with three large red blotches; head with several pearl-colored streaks (yellow in life); a transverse one between the nostrils; two oblique ones running from orbit towards subopercle, and a broad one from angle of mouth to angle of preopercle. Some yellow spots on sides of head. Each scale on sides with a yellow spot on its edge. Fins mostly red, the soft dorsal and anal with four rounded yellow spots; several spots on spinous dorsal and caudal (*Poey*). Eye rather large, as wide as interorbital space, shorter than snout. Maxillary reaching a little beyond eye. Edge of preopercle minutely denticulated, the angle rounded, projecting somewhat beyond the posterior edge; opercle with a membranaceous flap. Ventrals not reaching vent; caudal emarginate. Head 4 in total length; depth 4⅗. D. XI, 10; A. III, 10. Scales 2½–30–8. L. 10 inches. West Indies, north to Pensacola.

(*Cossyphus puellaris* Poey, Memorias Cuba, 1860, II, 210; Günther, IV, 101. Jordan, Proc. U. S. Nat. Mus., 1884.)

[2] *Platyglossus radiatus. Pudding-wife; Doncella; Blue-fish.*

This species (*Platyglossus radiatus* of the text; and *cyanostigma* of the addenda) is the original *Labrus radiatus* L., Syst. Nat., Ed. X, 288, 1758, based on *Turdus oculo radiato*, the Pudding-wife, of Catesby. It reaches a much larger size than our other species. The ground color in the males is blue, in the females chiefly of a bronze-olive. Both are most brilliantly colored. Lower pharyngeals T-shaped, but little broader than long.

[3] *Platyglossus bivittatus. Slippery Dick.*

This is the *Sparus radiatus* of Linnæus, Syst. Nat., Ed. XII, 472, 1766, based on a specimen sent from Charleston by Dr. Garden. It varies considerably with age and surroundings. The names *grandisquamis*, *humeralis*, and *florealis* represent different stages of growth. Lower pharyngeal T-shaped, more than twice as broad as long.

1161. **Platyglossus maculipinna**[1] Müller & Troschel. W.
1162. **Platyglossus semicinctus** Ayres. C. (949)
1163. **Platyglossus dispilus**[2] Günther. P.

372.—PSEUDOJULIS Bleeker. (320)

§ *Pseudojulis.*

1164. **Pseudojulis notospilus**[3] Günther. P.

§ *Oxyjulis.* Gill.

1165. **Pseudojulis modestus** Girard. C. (950)

373.—THALASSOMA[4] Swainson.

1166. **Thalassoma lucasanum** Gill. P.

374.—DORATONOTUS[5] Günther.

1167. **Doratonotus thalassinus** Jordan & Gilbert. W.

---

[1] *Platyglossus maculipinna* Müller & Troschel.
Dorsal fin with a black (blue) spot between the fifth and seventh spines and with a band along the middle of the soft portion; a small black spot posteriorly in the axil of the dorsal; a broad dark band runs from the head to the caudal fin, below the lateral line; sometimes a dark spot below the band on the middle of the body; a blue band from the snout through the eye to the operculum, and another above it from the snout to the eye; both bands are united, forming a V. Three bluish bands across the nape and three white ones on the cheek. Base of the pectoral with a small black spot. Caudal rounded. D. IX, 11; A. III, 11. Scales 2-28-9 (*Günther*), West Indies; a young specimen taken by us at Beaufort, N. C., in 1877.
(*Julis maculipinna* Müller & Troschel, Hist. Barbadoes, 674; Günther, IV, 165. "*Pusa*"? *radiata* Jor. & Gill., Proc. U. S. Nat. Mus. 1878, 374.)

[2] *Platyglossus dispilus* Günther, Proc. Zoöl. Soc. London, 1864, 25, and Fish. Centr. Amer., 1869, 417. Mazatlan to Panama.

[3] *Pseudojulis notospilus* Günther ll. cc. 26, 447. Mazatlan to Panama.

[4] THALASSOMA Swainson.

(*Julis* Günther, not of Cuvier, whose type *Labrus julis* L. is a species of *Coris*; not of Swainson, who also restricted *Julis* to the species of *Coris*.)
(Swainson, Classn. Anim. II, 1839, 224; type *Julis purpureus* Rüppell.)
This genus differs from *Platyglossus* in the possession of but eight spines in the dorsal, and in having no posterior canine tooth. The numerous species are gaily colored, like those of *Platyglossus*. They are found chiefly in the Western Pacific. (Θάλασσα, the sea; σῶμα, body, from the sea-green color of *T. purpureum*.)
Thalassoma lucasanum = *Julis lucasana* Gill., Proc. Ac. Nat. Sci. Phila., 1862, 142; *Julis lucasana* Günther, IV, 184. Gulf of California.

[5] DORATONOTUS Günther.

(Günther, Cat. Fishes Brit. Mus. IV, 124, 1862; type *Doratonotus megalepis* Günther.)
Body compressed; head not compressed to an edge anteriorly; its profile in front straight or concave; preorbital not very deep; mouth rather wide; teeth in a single series, two large canines in front in each jaw; a posterior canine; cheeks and opercles scaly; gill membranes united, free from the isthmus; scales large; lateral line interrupted behind, beginning again lower down; dorsal fin with nine strong pungent spines; some of the anterior elevated, the median spines short, so that the outline of the fin is concave; caudal rounded. Colors brilliant. Size small. Two species, each known from a single specimen. (Δόρυ (δορατος), spear; νῶτος, back.)
Doratonotus thalassinus Jordan & Gilbert, Proc., U. S. Nat. Mus., 1884, 28. Key West.

375.—XYRICHTHYS Cuvier. (321)

§ *Xyrichthys.*

1168. Xyrichthys psittacus[1] L. S. W. (951)
1169. Xyrichthys mundiceps[2] Gill. P.

§ *Iniistius* Gill.

1170. Xyrichthys mundicorpus[3] Gill. P.

§ *Dimalacocentrus* Gill.

1171. Xyrichthys rosipes[4] Jordan & Gilbert. W.

376.—CRYPTOTOMUS[5] Cope. (322)

1172. Cryptotomus ustus Cuv. & Val. W. (953)
1173. Cryptotomus beryllinus[6] Jordan & Swain. W.

377.—SPARISOMA[7] Swainson.

1174. Sparisoma radians Cuv. & Val. W. (954 d.)

---

[1] *Coryphæna psittacus* L., Syst. Nat., XII, 448, 1766 = *Coryphæna lineata* Gmelin = *Xyrichthys vermiculatus* Poey. The type of *Coryphæna psittacus* was sent from Charleston by Dr. Garden, and it has been identified as a *Xyrichthys* by Dr. Bean, who has examined it in London. Possibly another species of this type (*Xyrichthys venustus* Poey = *X. lineatus* C. & V.) occurs with the preceding on our coasts.

[2] *Xyrichthys mundiceps* Gill, Proc. Ac. Nat. Sci. Phila., 1862, 143; Jordan & Gilbert, Proc. U. S. Nat. Mus., 1882, 367. Cape San Lucas.

[3] *Iniistius mundicorpus* Gill, l. c., 1862, 145; *Novacula mundicorpus* Jordan & Gilbert, l. c., 367. Cape San Lucas. The subgenus, *Iniistius* (Gill, Proc. Ac. Nat. Sci. Phila., 1862, 145; type *Xyrichthys paro* Cuv. & Val.) is distinguished from *Xyrichthys* by the prolongation and separation from the fin of the first two dorsal spines.

[4] *Xyrichthys rosipes* Jordan & Gilbert, Proc. U. S. Nat. Mus., 1884, 27. Key West. The subgenus *Dimalococentrus* Gill (Proc. Ac. Nat. Sci. Phila., 1863, 223; type *Novaculichthys callosoma* Bleeker), is distinguished from *Xyrichthys* by the rounded (not trenchant) anterior edge of the head, and by the partial separation of the first two dorsal spines from the rest of the fin.

[5] *Cryptotomus* Cope (Trans. Am. Phil. Soc., 1871, 462; type *Cr. roseus* Cope) = *Calliodon* Cuv.; not of Bloch & Schneider, which is *Scarus* Forskål. For a detailed account of our genera and species of *Scaroid* fishes, see Jordan & Swain, Proc. U. S. Nat. Mus., 1884, 81.

[6] *Cryptotomus beryllinus* Jordan & Swain, Proc. U. S. Nat. Mus., 1884, 101. Key West and Havana.

[7] SCARUS Forskål.

The two groups *Scarus* (= *Hemistoma* Swainson, and *Pseudoscarus* Bleeker) and *Sparisoma* (= *Scarus* Bleeker) are really very distinct genera, each represented by several species among the Florida Keys. They may be thus defined:

SCARUS Forskål.

(*Calliodon* Gronow; *Hemistoma* Swainson; *Pseudoscarus* Bleeker.)

(Forskål, Descr. Anim. Orientali Observ., 1775, 25; type *Scarus psittacus* Forskål, &c.)

Lower pharyngeal spoon-shaped, much longer than broad, transversely concave; teeth fully coalesced, divided in each jaw by a distinct median suture; skull broad above; gill membranes forming a fold across the narrow isthmus; dorsal spines flex-

1175. **Sparisoma xystrodon**[1] Jordan & Swain. W.
1176. **Sparisoma cyanolene**[2] Jordan & Swain. W.
1177. **Sparisoma flavescens**[3] Bloch & Schneider. W. (954 c.)

378.—SCARUS Forskål. (323)

1178. **Scarus croicensis** Bloch. W. (954 b.)
1179. **Scarus cœruleus**[4] Bloch W.
1180. **Scarus guacamaia** Cuvier. W. (954)
1181. **Scarus perrico**[5] Jordan & Gilbert. P.

Family CXIII.—CICHLIDÆ. (95)

379.—HEROS Heckel. (324)

1182. **Heros cyanoguttatus** Baird & Girard. Vsw. (955)
1183. **Heros pavonaceus** Garman. Vsw. (955 b.)

Family CXIV.—POMACENTRIDÆ. (96)

380.—POMACENTRUS Lacépède.

§ *Pomacentrus.*

1184. **Pomacentrus obscuratus**[6] Poey. W.
1185. **Pomacentrus leucostictus** Müller & Troschel. W. (956)
1186. **Pomacentrus caudalis**[7] Poey. W.

---

ible, lateral line interrupted, its pores nearly simple; scales about head comparatively numerous, lower jaw included; upper pharyngeal teeth in two rows. Species mostly of large size, brightly colored; sexes similar.

SPARISOMA Swainson.
(*Scarus* Bleeker.)

(Swainson, Nat. Hist. Class'n Fishes, &c., 1839, II, 227; type *Sparus abildgaardii* Bloch.)

Lower pharyngeal much broader than long, its surface slightly concave; teeth less perfectly coalescent than in *Scarus*; the median suture not very distinct; skull narrow; gill membranes broadly united to the isthmus; dorsal spines pungent; lateral line continuous, its pores very much branched; scales about head few and large, those on cheeks in one row; lower jaw projecting; upper pharyngeal teeth in three rows. Species mostly of small size. (*Sparus*; σωμα, body.)

[1] *Sparisoma xystrodon* Jordan & Swain, l. c. 99. Havana and Key West.
[2] *Sparisoma cyanolene* Jordan & Swain, l. c. 98. Key West.
[3] For synonymy and description of *Sparisoma flavescens* (*Scarus squalidus* Poey), see Jordan & Swain, l. c. 92. Key West, southward.
[4] For synonymy and description of *Scarus cœruleus*, see Jordan & Swain, l. c. 85.
[5] *Scarus perrico* Jordan & Gilbert, Proc. U. S. Nat. Mus., 1881, 357. Mazatlan to Panama.
[6] *Pomacentrus obscuratus* Poey, Enumeratio Piscium Cubensium, 1875, 101; Jordan, Proc. U. S. Nat. Mus., 1884, 133. Key West to Cuba.
[7] *Pomacentrus caudalis* Poey, Synopsis Piscium Cubensium, 328, 1868.

Upper parts dusky, the greater part of each scale light grayish blue; lower parts bright yellow, with some blue spots on the scales; top and sides of head similarly marked with bluish spots on the scales. A jet-black, ink-like spot ocellated with blue on the back of the tail. Dorsal fin colored like the back; the posterior rays abruptly yellow; caudal fin bright yellow; lower fins chiefly yellow. Form oblong, ovate; the anterior profile moderately convex. Preorbital and preopercle well serrated. Teeth moderate, entire. Soft parts of dorsal and anal rather high. Head $3\frac{1}{4}$; depth $2\frac{1}{4}$. D. XII,14; A. II, 13. Scales 4-29-9. Cuba; lately obtained at Pensacola, by Silas Stearns.

1187. Pomacentrus rectifrænum [1] Gill. P.
1188. Pomacentrus flavilatus [2] Gill. P.

§ *Hypsypops* Gill.

1189. Pomacentrus quadrigutta [3] Gill. P.
1190. Pomacentrus rubicundus [4] Girard. C. (957)

381.—GLYPHIDODON Lacépède. (325 b.)

1191. Glyphidodon declivifrons Gill. W. P. (958)
1192. Glyphidodon saxatilis Linnæus. W. (959)
1192b. *Glyphidodon saxatilis troscheli* [5] Gill. P.

382.—CHROMIS Cuvier. (326)

1193. Chromis punctipinnis Cooper. C. (960)
1194. Chromis atrilobatus [6] Gill. P.
1195. Chromis insolatus Cuv. & Val. W. (961)
1196. Chromis enchrysurus Jordan & Gilbert. W. (961 b.)

Family CXV.—EPHIPPIDÆ. (97)

383.—CHÆTODIPTERUS Lacépède. (327)

1197. Chætodipterus faber Broussonet. N. S. W. (962)
1198. Chætodipterus zonatus [7] Girard. P.

Family CXVI.—CHÆTODONTIDÆ. (98)

384.—CHÆTODON Linnæus. (328)

1199. Chætodon maculocinctus Gill. (Acc.) (963)
1200. Chætodon ocellatus [8] Bloch. W. (963 b.)
1201. Chætodon capistratus Linnæus. W. (963 c.)
1202. Chætodon humeralis [9] Günther. P.
1203. Chætodon nigrirostris [10] Gill. P.

---

[1] *Pomacentrus rectifrænum* Gill, Proc. Ac. Nat. Sci., Phila. 1862, 148; 1863, 244 = *Pomacentrus analigutta* Gill, in Günther, IV, 27. Gulf of California to Panama.

[2] *Pomacentrus flavilatus* Gill, Proc. Ac. Nat. Sci. Phila., 1862, 148; 1863, 214 = *Pomatopricon bairdi* Gill, l. c., 1863, 217. Cape San Lucas. See Jordan & Gilbert, Proc. U. S. Nat. Mus., 1882, 365.

[3] *Hypsypops dorsalis* Gill, Proc. Ac. Nat. Sci. Phila. 1862, 147 = *Pomacentrus quadrigutta* Gill, Proc. Ac. Nat. Sci. Phila., 1862, 149; the name *dorsalis* is preoccupied in *Pomacentrus*. Cape San Lucas.

[4] For description of the young of *Pomacentrus rubicundus*, see Rosa Smith, Proc. U. S. Nat. Mus., 1882, 652.

[5] *Glyphidodon troscheli* Gill, Proc. Ac. Nat. Sci. Phila., 1862, 150. Cape San Lucas to Panama; perhaps not at all different from *G. saxatilis*.

[6] *Chromis atrilobatus* Gill, Proc. Ac. Nat. Sci. Phila., 1862, 149. Cape San Lucas to Panama.

[7] *Ephippus zonatus* Girard, U. S. Pac. R. R. Ex pl., 1858, 110. San Diego to Panama. Pacific coast specimens of *Chætodipterus* differ from the ordinary *C. faber* in the less development of the third dorsal spine, which is little longer or higher than the others. The dark bands are usually more obscure in *C. zonatus*. In other respects the two forms agree very closely.

[8] *Chætodon ocellatus* Bloch, Ichth. tab. 211 = *Chætodon bimaculatus* Bloch, tab. 219. See Poey, Enum. Pisc. Cubens., 1875, 62.

[9] *Chætodon humeralis* Günther, II, 19, 1860. Mazatlan to Panama.

[10] *Sarothrodus nigrirostris* Gill, Proc. Ac. Nat. Sci. Phila., 1862, 243. Cape San Lucas.

385.—HOLACANTHUS Lacépède.

1204. **Holacanthus strigatus**[1] Gill. P.
1205. **Holacanthus ciliaris** Linnæus. W. (964)

386.—POMACANTHUS Lacépède. (329)

§ *Pomacanthodes* Gill.
1206. **Pomacanthus zonipectus**[2] Gill. P.

§ *Pomacanthus.*
1207. **Pomacanthus aureus**[3] Bloch. W.

Family CXVII.—ACANTHURIDÆ. (99)

387.—TEUTHIS[4] Linnæus. (330)

1208. **Teuthis hepatus** Linnæus. S. W. (966)
1209. **Teuthis tractus** Poey. W. P. (966 c.)
1210. **Teuthis cœruleus** Bloch. W. (967)

388.—PRIONURUS[5] Lacépède.

1211. **Prionurus punctatus** Gill. P.

---

[1] *Holacanthus strigatus* Gill, Proc. Ac. Nat. Sci. Phila., 1862, 243. Cape San Lucas to Panama. *Holacanthus tricolor* (Synopsis, p. 941) should be omitted. It has not yet been taken at the Florida Keys, although doubtless occurring there.

[2] *Pomacanthodes zonipectus* Gill, Proc. Ac. Nat. Sci. Phila., 1862, 244 (adult) = *Pomacanthus crescentalis* Jordan & Gilbert, Proc. U. S. Nat. Mus., 1881, 358 (Young). Gulf of California to Panama.

[3] *Pomacanthus aureus* (Bloch), *Black Angel, Chirivita*. The description of *Pomacanthus arcuatus*, on page 616 of the Synopsis, was taken from a specimen of this species, with the exception of the following phrases, which should be suppressed: "Young with yellowish vertical bands"; the bands in the young of *P. aureus* are whitish. "Lat. l. 80-100"; this should read, "lat. l. 65." The additional characters given on page 973 are taken from the true *P. arcuatus*, and should be suppressed, as should also the synonymy on page 616. The true *arcuatus* is a West Indian species, not yet known from our coast; it is darker and more uniform in color than *P. aureus*, the cross bands in the young are better defined and are yellow; the scales are smaller (lat. l. 85 to 90); and the dorsal spines are almost invariably 10 instead of 9. *P. aureus* is common in the West Indies and north to the Florida keys.

(*Chætodon aureus* Bloch, Ichthyol.; tab. 193, f. l.; Cuvier & Val., VII, 202, 1831; *Pomacanthus balteatus* and *arcuatus* Cuv. & Val., VII. 208, 211; *Chætodon aureus* Poey, Syn. Pisc., Cubens., 1875, 60; *Chætodon aureus* Bleeker, Archives Neerlandaises, IX, 1876, 183; Lütken, Spolia Atlantica, 1880, 571.)

[4] The genus *Teuthis* of Linnæus, Systema Naturæ, is based on *Teuthis hepatus* L. This species, founded on *Hepatus* of Gronow, is the common species known as *Acanthurus chirurgus*, with which *A. phlebotomus* Cuv. & Val. (*nigricans* of the Synopsis) seems to be identical. The generic name *Acanthurus* must give place to *Teuthis*, and this species should stand as *Teuthis hepatus*. See Gill, Proc. Ac. Nat. Mus., 1884, 275, and Meek and Hoffman, Proc. Ac. Nat. Sci. Phila., 1884. In the latter paper is given a detailed account of the three American species of *Teuthis*.

[5] PRIONURUS Lacépède.

(Lacépède, Annales Museum, Paris, IV, 205; type *Prionurus microlepidotus* Lac.) This genus differs from *Teuthis* chiefly in the armature of the tail, which consists of a series of 3 to 6 bony keeled laminæ on each side. Size small. Species not very numerous, in the tropical seas. ($\Pi\rho\iota\omega\nu$, saw ; $o\upsilon\rho\alpha$, tail.)

Prionurus punctatus Gill, Proc. Ac. Nat. Sci. Phila., 1862, 242. Cape San Lucas.

Family CXVIII.—TRACHYPTERIDÆ. (100)

389.—TRACHYPTERUS Gouan. (331)

1212. Trachypterus altivelis Kner. B. C. (968)

Family CXIX.—BATHYMASTERIDÆ.[1]

390.—BATHYMASTER Cope. (334)

1213. Bathymaster signatus Cope. A. (971)

Family CXX.—MALACANTHIDÆ. (102)

391.—LOPHOLATILUS Goode & Bean. (335)

1214. Lopholatilus chamæleonticeps Goode & Bean. B. (972)

392.—CAULOLATILUS Gill. (336)

1215. Caulolatilus princeps Jenyns. C. P. (973)
1216. Caulolatilus microps[2] Goode & Bean. W. (974)

Family CXXI.—GOBIIDÆ. (104)

393.—GOBIOMORUS Lacépède. (339)

1217. Gobiomorus dormitator Lacépède. W. Vsw. (978)
1218. Gobiomorus lateralis Gill.[3] P.

394.—EROTELIS Poey.

1219. Erotelis smaragdus[4] Cuv. & Val. W.

---

[1] I have here dismembered the unnatural group of *Icosteidæ* as given in the Synopsis, referring *Icosteus* and *Icichthys*, in accordance with the views of Dr. Steindachner (Ichth. Beitr., XI, 4, 1881, and XII, 22, 1882), to the Scombroid series, in the neighborhood of the *Bramidæ*. Steindachner considers *Schedophilus* the nearest ally of *Icosteus* (= *Schedophilopsis spinosus* Steindachner l. c.), and this may be correct.

The genus *Bathymaster* is perhaps the type of a separate family, allied to *Malacanthus*, *Latilus*, &c., or perhaps to *Opisthognathus*. For the present, I unite the *Latilidæ* with the *Malacanthidæ*, leaving *Bathymaster* in a group by itself. This arrangement is, however, merely provisional, until the anatomy of the different forms is made known.

[2] *Caulolatilus microps* Goode & Bean.

The identity of our Atlantic species of *Caulolatilus* with either the Cuban *cyanops* or the Brazilian *chrysops* is as yet unproven, though not improbable. The scales in our species are smaller than they are said to be in the others. There is little difference between *C. microps* and *C. princeps* except in color. The scales of the body have each a small brownish spot at base in *C. microps*.

[3] *Philypnus lateralis* Gill, Proc. Ac. Nat. Sci. Phila., 1860, 123; Jordan & Gilbert, Proc. U. S. Nat. Mus., 1882, 380. Streams of Northwestern Mexico.

[4] *Eleotris smaragdus* Cuv. & Val. *Esmeralda negra*.

Dusky olive, the fins mostly bluish, the dorsal with brown lines; some dark markings about eye, and on base of pectoral above. Body very long and slender, compressed behind, the form much as in *Gobionellus oceanicus*. Head depressed, flattish above, the eyes mostly superior, not half the width of the interorbital area, which has a knob near its middle. Mouth very oblique, the lower jaw much projecting,

395.—ELEOTRIS (Gronow) Bloch & Schneider. (340, 341 b.)

1220. **Eleotris pisonis** Gmelin. W. (981)
1221. **Eleotris amblyopsis** Cope. S. W. (981 b.)
1222. **Eleotris æquidens**[1] Jordan & Gilbert. P.

396.—DORMITATOR Gill. (341)

1223. **Dormitator maculatus** Bloch. W. (980, 981)
1224. **Dormitator latifrons**[2] Richardson. P.

397.—GOBIUS Linnæus.

§ *Euctenogobius* Gill.

1225. **Gobius lyricus** Girard. S. (983)
1226. **Gobius encæomus** Jordan & Gilbert. S. (983 b.)

§ *Rhinogobius* Gill.

1227. **Gobius banana**[3] Cuv. & Val. P. W.

§ *Gobius*.

1228. **Gobius soporator** Cuv. & Val. S. W. P. (984, 982, 985)

§ *Coryphopterus* Gill.

1229. **Gobius sagittula**[4] Günther. P.
1230. **Gobius boleosoma** Jordan & Gilbert. S. (987 b.)
1231. **Gobius stigmaturus** Goode & Bean. S. (987 c.)
1232. **Gobius würdemanni**[5] Girard. S. (987)
1233. **Gobius nicholsi** Bean. A. (987 d.)
1234. **Gobius glaucofrænum** Gill. A. (988)

---

the maxillary about reaching front of eyes; teeth rather small, in bands. Fins rather high; dorsal spines slender, lower than the highest soft rays, which are 1¼ in head. Caudal lanceolate, ⅓ longer than head. Ventrals moderate, 2 in head. Scales very small cycloid. Head 5¼; depth 10 to 12 D. VI-I, 10. A, I, 9. Lat. l. about 100. L. 8 inches. West Indies, north to Key West, not ascending the fresh waters.
(Cuv. & Val., XII, 231, 1837; *Erotelis valenciennesi* Poey, Mem. Cuba, II, 273, 1860. Günther, III, 123.)

This species is the type of Poey's genus *Erotelis* (name an anagram of *Eleotris*), distinguished from *Eleotris* by the very slender form, similar to that of *Gobionellus*.

[1] *Culius æquidens* Jordan & Gilbert, Proc. U. S. Nat. Mus., 1881, 461. Fresh waters of Western Mexico and Lower California.

[2] *Eleotris latifrons* Richardson, Voyage Sulphur, Fishes, 57 = *Dormitator microphthalmus* Gill. Streams of the Pacific coast, north to Lower California. There are some tangible differences between the specimens of *Dormitator* found on the west coast of Mexico and that found in the Atlantic waters. For an excellent account of the genera and species of *Eleotridinæ*, see Eigenmann and Fordise, Proc. Ac. Nat. Sci. Phila., 1885.

[3] *Gobius banana* Cuv. & Val., XII, 103; Günther, III, 59; Jordan & Gilbert, Proc. U. S. Nat. Mus., 1882, 379. Tropical America, north to Lower California, in fresh water.

[4] *Euctenogobius sagittula* Günther, III, 555. *Gobius sagittula* Jordan & Gilbert, Proc. U. S. Nat. Mus., 1882, 380. Lower California to Panama.

[5] For description of *Gobius würdemanni* see Jordan, Proc. U. S. Nat. Mus., 1884, 321.

398.—GOBIONELLUS Girard. (345)

1235. Gobionellus oceanicus Pallas. S. W. (989)
1236. Gobionellus stigmaticus Poey. W. (989 b.)

399.—GILLICHTHYS Cooper. (346)

1237. Gillichthys mirabilis Cooper. C. (990)

400.—LEPIDOGOBIUS Gill. (347)

§ *Lepidogobius* Gill.

1238. Lepidogobius lepidus Girard. C. (991)

§ *Encyclogobius* Gill.

1239. Lepidogobius newberryi Girard. C. (992)
1240. Lepidogobius gulosus Girard. S. (992 b; 986)
1241. Lepidogobius thalassinus Jordan & Gilbert. S. (992 b.)

401.—GOBIOSOMA[1] Girard. (348)

1242. Gobiosoma ceuthœcum Jordan & Gilbert. W.
1243. Gobiosoma bosci Lacépède. N. S. (993; 994)
1244. Gobiosoma histrio[2] Jordan. P.
1245. Gobiosoma zosterurum[3] Jordan and Gilbert. P.
1246. Gobiosoma longipinne[4] Steindachner. P.
1247. Gobiosoma ios Jordan & Gilbert. C. (994 b.)

402.—TYPHLOGOBIUS Steindachner. (349)

1248. Typhlogobius californiensis Steindachner. C. (995)

403.—TYNTLASTES Günther. (350)

1249. Tyntlastes sagitta Günther. P. (996)

404.—IOGLOSSUS Bean. (350 b.)

1250. Ioglossus calliurus Bean. S. (996 b.)

Family CXXII.—CHIRIDÆ. (105)

405.—PLEUROGRAMMUS Gill. (351 a.)

1251. Pleurogrammus monopterygius Pallas. A. (997)

406.—HEXAGRAMMUS Steller. (351 b.)

1252. Hexagrammus ordinatus Cope. A. (998.)
1253. Hexagrammus asper Steller. A. (999)

---

[1] *Gobiosoma ceuthœcum* Jordan & Gilbert, Proc. U. S. Nat. Mus., 1884, 29. Key West; found in the cavity of a sponge.

[2] *Gobiosoma histrio* Jordan, Proc. U. S. Nat. Mus., 1884, 260. Guaymas.

[3] *Gobiosoma zosterurum* Jordan & Gilbert, Proc. U. S. Nat. Mus., 1881, 361. Mazatlan.

[4] *Gobiosoma longipinne* Steindachner, Ichth. Beitr., VIII, 1879, 24. Las Animas, Gulf of California.

1254. **Hexagrammus scaber** Bean. A. (999 b.)
1255. **Hexagrammus superciliosus** Pallas. A. C. (1000)
1256. **Hexagrammus decagrammus** Pallas. A. C. (1001)

407.—OPHIODON Girard. (352)

1257. **Ophiodon elongatus** Girard. C. A. (1002)

408.—ZANIOLEPIS Girard. (353)

1258. **Zaniolepis latipinnis** Girard. C. (1003)

409.—OXYLEBIUS Gill. (354)

1259. **Oxylebius pictus** Gill. C. (1004)

410.—MYRIOLEPIS Lockington. (355)

1260. **Myriolepis zonifer** Lockington. C. (1005)

411.—ANOPLOPOMA Ayres. (356)

1261. **Anoplopoma fimbria** Pallas. C. A. (1006)

Family CXXIII.—SCORPÆNIDÆ. (106)

412.—SEBASTES Cuvier. (357)

1262. **Sebastes marinus** Linnæus. G. N. Eu. (1007)

413.—SEBASTODES Gill. (358)

1263. **Sebastodes paucispinis** Ayres. C. (1008)

414.—SEBASTICHTHYS Gill.

§ *Sebastosomus* Gill.

1264. **Sebastichthys flavidus** Ayres. C. (1009)
1265. **Sebastichthys melanops** Girard. C. (1010)
1266. **Sebastichthys ciliatus** Tilesius. A. (1011)
1267. **Sebastichthys mystinus** Jordan & Gilbert. C. (1012)
1268. **Sebastichthys entomelas** Jordan & Gilbert. C. (1013)
1269. **Sebastichthys ovalis** Ayres. C. (1014)
1270. **Sebastichthys proriger** Jordan & Gilbert. C. (1015)
1271. **Sebastichthys brevispinis**[1] Bean. A.
1272. **Sebastichthys atrovireus** Jordan & Gilbert. C. (1016)
1273. **Sebastichthys pinniger** Gill. C. (1017)

---

[1] *Sebastichthys brevispinis* (Bean). Closely allied to *S. proriger*, but larger in size and more uniform in color; second anal spine shorter than third; peritoneum white. Coast of Alaska. (Bean.)

(*Sebastichthys proriger* var. *brevispinis* Bean., Proc., U. S. Nat. Mus., 1883. *Sebastodes proriger*, Alaskan specimens, Jor. & Gilb., Syn. Fish. N. A., 1883, 950.)

The statement in the Synopsis, p. 950, that *S. proriger* has been confounded by Tilesius and Pallas with *S. ciliatus* is erroneous. The specimens called by them *ciliatus* and *variabilis* include *ciliatus* and *matzubaræ*. The true *proriger* is not yet known from Alaska.

1274. **Sebastichthys miniatus** Jordan & Gilbert. C. (1018)
1275. **Sebastichthys matzubaræ**[1] Hilgendorf. A.

§ *Sebastomus* Gill.

1276. **Sebastichthys ruber** Ayres. C. (1019)
1277. **Sebastichthys umbrosus** Jordan & Gilbert. C. (1019b.)
1278. **Sebastichthys constellatus** Jordan & Gilbert. C. (1020)
1279. **Sebastichthys rosaceus** Girard. C. (1021)
1280. **Sebastichthys rhodochloris** Jordan & Gilbert. C. (1022)
1281. **Sebastichthys chlorostictus** Jordan & Gilbert. C. (1023)
1282. **Sebastichthys elongatus** Ayres. C. (1024)
1283. **Sebastichthys rubrovinctus** Jordan & Gilbert. C. (1025)

§ *Sebastichthys.*

1284. **Sebastichthys auriculatus** Girard. C. (1026)
1285. **Sebastichthys rastrelliger** Jordan & Gilbert. C. (1027)
1286. **Sebastichthys caurinus** Richardson. A. (1028)
1286 b. *Sebastichthys caurinus vexillaris* Jordan & Gilbert. C. (1028 b.)
1287. **Sebastichthys maliger** Jordan & Gilbert. C. (1029)
1288. **Sebastichthys carnatus** Jordan & Gilbert. C. (1030)
1288 b. *Sebastichthys carnatus chrysomelas* Jordan & Gilbert. C. (1031)
1289. **Sebastichthys nebulosus** Ayres. C. (1032)
1290. **Sebastichthys serriceps** Jordan & Gilbert. C. (1033)
1291. **Sebastichthys nigrocinctus** Ayres. C. (1034)

415.—**SEBASTOPSIS**[2] Gill.

1292. **Sebastopsis xyris** Jordan & Gilbert. P.

416.—**SEBASTOPLUS**[3] Gill.

1293. **Sebastoplus dactylopterus** De la Roche. B. Eu. (1035)

---

[1] *Sebastichthys matzubaræ* (Hilgendorf). Dark red; three dark shades across cheeks. Allied to *Sebastichthys miniatus*. Spines of head low, developed about as in *S. miniatus* and *S. pinniger*. Preocular, supraocular, postocular, tympanic, occipital, and nuchal spines distinct; a pair of small coronal spines present, as also a small spine before and one just below eye. Maxillary reaching to posterior border of eye 1¾ in head. Both jaws covered with rough, ctenoid scales. Interorbital space flattish, scaled, its breath a little less than that of eye. Preopercular spine short, simple. Preorbital spines simple. Lower jaw scarcely projecting. Second anal spine scarcely longer than third. Longest dorsal spine 2¾ in head, a little less than the longest short rays. Pectoral 4½ in body.

Color chiefly red; three dark shades across cheek. D. XIII, 14. A. III, 7. Yeso; Aleutian Islands. The above description from a specimen in the Berlin Museum, brought by Pallas from the Aleutian Islands.

(*Perca variabilis* Pallas, Zoogr. Rosso. Asiat., III, 241, 1811, in part; the larger specimen, No. 8145, Berl. Mus.; *Sebastes matzubaræ* Hilgendorf, Sitzber. Gesellschaft Naturforschender Freunde, Berlin, 1880, 170; Jordan, Proc. Ac. Nat. Sci. Phila., 1883, 291.)

[2] SEBASTOPSIS Gill.

(Gill, Proc. Ac. Nat. Sci. Phila., 1862, 278; type *Sebastes polylepis* Bleeker.

This genus differs from *Sebastichthys* in the absence of palatine teeth. The known species are small in size and not very numerous. (*Sebastes*; ὄψις, appearance.)

Sebastopsis xyris Jordan & Gilbert, Proc. U. S. Nat. Mus., 1882, 369. Cape San Lucas.

[3] SEBASTOPLUS Gill.

(Gill, Proc. Ac. Nat. Sci. Phila., 1863, 207; type *Sebastes kuhli* Lowe.)

This genus includes species which have the general characters of *Sebastichthys*, with the vertebræ and dorsal spines in smaller number, as in *Scorpæna*.

The species are red in color and mostly inhabit deep water. (*Sebastes*; ὅπλος, armed.)

417.—SCORPÆNA Linnæus. (359)

1294. Scorpæna guttata Girard. C. (1036)
1295. Scorpæna plumieri Bloch. W. P. (1037)
1296. Scorpæna grandicornis[1] Cuv. & Val. W.
1297. Scorpæna brasiliensis[2] Cuv. & Val. W. S. (1038 b.)
1298. Scorpæna occipitalis[3] Poey. W. (1038 c.)

418.—SETARCHES Johnson. (360)

1299. Setarches parmatus Goode. B. (1039)

Family CXXIV.—COTTIDÆ. (107)

419.—HEMITRIPTERUS Cuvier.

1300. Hemitripterus americanus Gmelin. G. N. (1040)
1300 b. *Hemitripterus americanus cavifrons*[4] Lockington. A. (1041)

420.—ASCELICHTHYS Jordan & Gilbert. (362)

1301. Ascelichthys rhodorus Jordan & Gilbert. A. (1042)

421.—PSYCHROLUTES Günther. (363)

1302. Psychrolutes paradoxus Günther. A. (1043)

---

[1] *Scorpæna grandicornis* Cuv. & Val.
 Gray, with brown shades and faint cross-bars; sides with numerous bright yellow spots in life; axil dark gray, with round white dots, each surrounded by a dark ring. Pectoral largely blackish above; a black blotch at base below; the fin largely tinged with yellow, especially on the inner side. Supraocular filament blackish, with gray fringes. Soft dorsal largely blackish toward the tip; spinous dorsal chiefly dusky; ventrals tipped with blackish; anal with three black bands; caudal with two; a faint band at its base. Body rather stout; deeper than in *S. plumieri* and much less variegated in color. Sides and head with dermal flaps; a slight depression below eye; occipital pit very deep; spines of head sharp. A few scales on opercle. Breast with rudimentary scales. Supraocular flap very large, wide and fringed, more than half length of head, reaching to beyond front of dorsal. Maxillary reaching posterior margin of eye, $2\frac{1}{5}$ in head. Dorsal spines higher than in related species, the highest equal to second spine of anal and about half head. Head, $2\frac{1}{4}$; depth, $2\frac{1}{4}$. D. XII, 9. A. III, 5. Lat. l, 26 (pores.)
 West Indies, north to Key West.
 (Cuv. & Val., IV, 1829, 309; Günther, II, 115; Poey, Syn. Pisc. Cubens. 303.)
 The species of *Scorpæna* found in our waters may be readily distinguished by the color of the axillary region as follows:
 Guttata: pale, usually unspotted; one or two dark spots behind it.
 Plumieri: jet black, with a few large white spots.
 Brasiliensis: pale, with several round blackish spots.
 Occipitalis: pale, with dark specks, and a black spot above.
 Grandicornis: dusky gray, with numerous white stellate spots.
 [2] *Scorpæna brasiliensis* Cuv. & Val., V, 105; Günther, II, 312 = *Scorpæna stearnsi* Goode & Bean. South Carolina to Brazil.
 [3] *Scorpæna occipitalis* Poey, (Memorias Cuba, II, 171), is probably identical with *Scorpæna calcarata* Goode & Bean.
 [4] According to Dr. Bean, *Hemitripterus cavifrons* is not distinct from *H. americanus.*

422.—COTTUNCULUS Collett. (364)

1303. Cottunculus microps Collett. B. Eu. (1044)
1304. Cottunculus torvus[1] Goode. B. (1045).

423.—ARTEDIUS Girard.

1305. Artedius lateralis Girard. C. (1046)
1306. Artedius notospilotus Girard. C. (1047)
1307. Artedius fenestralis[2] Jordan & Gilbert. A. (365)

424.—ICELUS Kröyer.

1308. Icelus bicornis[3] Reinhardt. (1048, 1053, 1083)

425.—ICELINUS[4] Jordan.

1309. Icelinus quadriseriatus Lockington. C. (1049)

426.—CHITONOTUS Lockington.

1310. Chitonotus megacephalus Lockington. C. (1050)
1311. Chitonotus pugetensis Steindachner. A. (1051)

427.—ARTEDIELLUS[5] Jordan.

1312. Artediellus uncinatus Reinhardt. G. B. (1052)

428.—URANIDEA De Kay. (366)

Tauridea Jordan & Rice.

1313. Uranidea ricei Nelson. Vn. (1054)

Cottopsis Girard.

1314. Uranidea aspera Richardson. T. (1055)
1315. Uranidea semiscabra Cope. R. (1056)
1316. Uranidea rhothea Rosa Smith. T. (1056 b.)

---

[1] *Cottunculus torvus* is described in full by Goode, Bull. Mus. Comp. Zoöl., XIX, 212. Mr. Goode counts D. VII, 14; A. 13.

[2] *Artedius fenestralis* Jordan & Gilbert, Proc. U. S. Nat. Mus., 1882, 577. Puget Sound.

[3] According to Lütken (Vidensk. Meddels. naturh. Foren. Kjøb., 1876, 92), *Cottus bicornis* Reinhardt is identical with *Icelus hamatus* Kröyer. It is thought by Lütken that *Cottus polaris* Sabine is probably also the same fish, but if so, the description of Sabine is very erroneous. Nos. 1053 and 1083 may therefore be erased, and the species *Icelus hamatus* in the Synopsis may stand as *Icelus bicornis*.

[4] *Icelinus*, genus or subgenus nova for *Artedius quadriseriatus* Lockington, characterized by the peculiar squamation, preopercular armature, and form of the body as described in the Synopsis, p. 691. (Name a diminutive of *Icelus*.)

[5] ARTEDIELLUS Jordan.

(Genus nova; type *Cottus uncinatus* Reinhardt.)

This genus or subgenus differs from *Icelus* proper, apparently its nearest ally, in having the skin naked and smooth. *Centridermichthys* Richardson, an Asiatic genus to which this and other American species have been sometimes referred, has the skin prickly, and a large slit behind the fourth gill, the gill membranes being fully united to the isthmus. (A diminutive of *Artedius*.)

CATALOGUE OF THE FISHES OF NORTH AMERICA.

§ *Potamocottus* Gill.

1317. **Uranidea gulosa** Girard. T. (1057)
1318. **Uranidea punctulata** Gill. R. (1058)
1319. **Uranidea bendirei** Bean. R. (1059)
1320. **Uranidea richardsoni** Agassiz. V. (1060)
1320 b. *Uranidea richardsoni bairdi* Girard. Vnc.
1320 c. *Uranidea richardsoni kumlieni* Hoy. Vn.
1320 d. *Uranidea richardsoni wilsoni* Girard. Vn.
1320 e. *Uranidea richardsoni alvordi* Girard. Vn.
1320 f. *Uranidea richardsoni meridionalis* Girard. Ve.
1320 g. *Uranidea richardsoni zophera* Jordan. Vs.
1320 h. *Uranidea richardsoni carolinæ* Gill. Vs.
1320 i. *Uranidea richardsoni wheeleri* Cope. R.

§ *Uranidea*.

1321. **Uranidea cognata** Richardson. Vu. (1062)
1322. **Uranidea minuta** Pallas. Y. (1063)
1323. **Uranidea spilota**[1] Cope. Vn. (1062 b.)
1324. **Uranidea pollicaris** Jordan & Gilbert. Vu. (1062 o.)
1325. **Uranidea marginata** Bean. R. (1064)
1326. **Uranidea viscosa** Haldeman. Ve. (1065)
1327. **Uranidea gracilis** Heckel. Vc. (1066)
1328. **Uranidea gobioides** Girard. Ve. (1067)
1329. **Uranidea boleoides** Girard. Ve. (1068)
1330. **Uranidea franklini** Agassiz. Vn. (1069)
1331. **Uranidea formosa** Girard. Vn. (1069 b.)
1332. **Uranidea hoyi** Putnam. Vn. (1070)

429.—COTTUS Linnæus. (367)

1333. **Cottus octodecimspinosus**[2] Mitchill. N. (1072)
1334. **Cottus æneus** Mitchill. N. (1073)
1335. **Cottus scorpioides** Fabricius. G. (1074)
1336. **Cottus scorpius** L. G. Eu. (1075)
1336 b. *Cottus scorpius grönlandicus* Cuv. & Val. N. G. (1075 b.)
1337. **Cottus polyacanthocephalus**[3] Pallas. A. (1076, 1081)
1338. **Cottus labradoricus** Girard. G. (1077)
1339. **Cottus tæniopterus** Kner. A. (1078)
1340. **Cottus quadricornis** L. G. Eu. (1079)
1341. **Cottus humilis** Bean. A. (1080)
1342. **Cottus axillaris** Gill. A. (1082)
1343. **Cottus platycephalus**[4] Pallas. A. (1084)
1344. **Cottus verrucosus** Bean. A. (1085)
1345. **Cottus niger** Bean. A. (1086)
1346. **Cottus quadrifilis** Gill. A. (1087)

---

[1] I have re-examined the type of *Uranidea spilota*. It has now no evident teeth on the palatines and the ventral rays are I, 3. The skin is smooth, and the preopercular spine, although prominent and directed upward, is not hooked. The spots on the body are less sharply defined than in *U. ricei*.

[2] *Cottus bubalis* should be omitted. It is a European species, and it has not yet been found in Greenland, according to Dr. Lütken.

[3] *Cottus jaok* should be omitted. The type, lately examined by Dr. Bean in Berlin, is identical with *Cottus polyacanthocephalus*.

[4] *Cottus platycephalus* Pallas, the type of which has been lately re-examined by Dr. Bean and the writer, is a valid species of *Cottus*. It has no palatine teeth.

430.—GYMNACANTHUS Swainson. (368)

1347. Gymnacanthus tricuspis[1] Reinhardt. G.
1348. Gymnacanthus pistilliger Pallas. A. (1088)
1349. Gymnacanthus galeatus Bean. A. (1089)

431.—TRIGLOPSIS Girard. (369)

1350. Triglopsis thompsoni Girard. Vn. (1090)

432.—ENOPHRYS Swainson. (370)

1351. Enophrys bison Girard. C. A. (1091)
1352. Enophrys diceraus[2] Pallas. A. (1092, 1093)

433.—LIOCOTTUS Girard. (371)

1353. Liocottus hirundo Girard. C. (1094)

434.—TRIGLOPS Reinhardt. (372)

1354. Triglops pingeli Reinhardt. G. Eu. A. (1095)

435.—PRIONISTIUS[3] Bean.

1355. Prionistius macellus Bean. A.

436.—LEPTOCOTTUS Girard. (373)

1356. Leptocottus armatus Girard. C. (1096)

437.—HEMILEPIDOTUS Cuvier. (374)

1357. Hemilepidotus spinosus Ayres. C. (1097)
1358. Hemilepidotus jordani Bean. A. (1098)
1359. Hemilepidotus hemilepidotus Tilesius. A. (1099)

438.—MELLETES Bean. (375)

1360. Melletes papilio Bean. A. (1100)

439.—SCORPÆNICHTHYS Girard. (376)

1361. Scorpænichthys marmoratus Ayres. C. (1101)

---

[1] Mr. Dresel observes (Proc. U. S. Nat. Mus., 1884, 251): Dr. T. H. Bean "inclines to the belief that the Greenland form of *Gymnacanthus* (*tricuspis*) does not occur in the Pacific. It is best, therefore, to retain Reinhardt's name, *tricuspis*, for the Atlantic species." A description of *G. tricuspis* is given by Mr. Dresel, l. c. The description in the Synopsis is also from an Atlantic specimen.

[2] *Enophrys claviger* is the young of *E. diceraus*, according to Dr. Bean, who has examined the types of both species.

[3] PRIONISTIUS Bean.

(Bean, Proc. U. S. Nat. Mus., 1883, 355; type *Prionistius macellus* Bean.)

Allied to *Triglops*, differing in the following respects: the much slenderer form; the absence of a series of bony tubercles along the bases of the dorsal fins, the elongation of the exserted pectoral rays so that the lower portion of the fin is considerably longer than the upper, the presence of serrations on all the dorsal spines and on the first soft ray, and the emargination of the caudal fin. Alaska. ($\Pi\rho\iota o\nu$, saw; $\iota\sigma\tau\iota o\nu$, sail; dorsal fin.)

Prionistius macellus Bean, l. c. Coast of British Columbia.

440.—OLIGOCOTTUS Girard. (377)

§ *Clinocottus* Gill.

1362. Oligocottus analis Girard. C. (1102)

§ *Oligocottus*.

1363. Oligocottus maculosus Girard. C. (1103)

§ *Blennicottus* Gill.

1364. Oligocottus globiceps Girard. C. (1104)

441.—BLEPSIAS Cuvier. (378)

1365. Blepsias cirrhosus Pallas. A. (1105)
1366. Blepsias bilobus Cuv. & Val. A. (1106)

442.—NAUTICHTHYS Girard. (379)

1367. Nautichthys oculofasciatus Girard. A. (1107)

443.—RHAMPHOCOTTUS Günther. (380)

1368. Rhamphocottus richardsoni Günther. A. (1108)

Family CXXV—AGONIDÆ (108 *a*.)

444.—ASPIDOPHOROIDES Lacépède. (381)

1369. Aspidophoroides monopterygius Bloch. N. G. (1109)
1370. Aspidophoroides inermis Günther. A. (1110)
1371. Aspidophoroides olriki[1] Lütken. G.
1372. Aspidophoroides güntheri Bean. A.

445.—SIPHAGONUS Steindachner. (382)

1373. Siphagonus barbatus Steindachner. G. (1111)

446.—BRACHYOPSIS[2] Gill. (383)

1374. Brachyopsis rostratus Tilesius. A. (1112)

---

[1] *Aspidophoroides olriki* Lütken.
Body short and thick, much less elongate than in the other species of this genus; head broad, the interorbital space concave, as is the median line of the back; lower jaw included; snout with a short spine above; no barbels; shields without spines; breast with about ten conical striate shields. Fins very much larger than in the other species of *Aspidophoroides*, the dorsal fin about as high as long, but little larger than anal. Ventrals small, 2⅜ in head; pectorals about as long as head. Head 4⅞; depth 6. D. 6 or 7. A. 6 or 7. V. 1, 2. P. 13. C. 10. L. 4 inches. Greenland. from the stomachs of flounders.
(Lütken, Nordiske Ulketiske, Vidensk. Meddels. naturh. Foren., Kjöbenhavn, 1876, 385.)

[2] The name *Brachyopsis* should be retained for this genus, instead of *Leptagonus*. "*Leptagonus*" *decagonus*, lately examined by me in Copenhagen, has the gill membranes attached to the isthmus and forming a narrow fold across it. It should, therefore, be referred to *Podothecus*, although in some respects approaching *Agonus*, rendering a reunion of these genera probably necessary.

1375. **Brachyopsis verrucosus** Lockington. C. (1113)
1376. **Brachyopsis xyosternus** Jordan & Gilbert. C. (1114)

447.—BOTHRAGONUS Gill. (385)

1377. **Bothragonus swani** Steindachner. A. (1117)

448.—ODONTOPYXIS Lockington. (386)

1378. **Odontopyxis trispinosus** Lockington. C. (1118)

449.—PODOTHECUS Gill. (387)

§ *Leptagonus* Gill.
1379. **Podothecus decagonus** Bloch & Schneider. G. (1115)

§ *Podothecus.*
1380. **Podothecus vulsus** Jordan & Gilbert. C. (1119)
1381. **Podothecus acipenserinus** Tilesius. A. (1120)

Family CXXVI.—TRIGLIDÆ. (108 b.)

450.—PERISTEDION Lacépède. (388)

1382. **Peristedium miniatum.** Goode. B. (1121)
1383. **Peristedium imberbe**[1] Poey. W. B.

451.—PRIONOTUS Lacépède. (390)

§ *Ornichthys* Swainson.
1384. **Prionotus scitulus**[2] Jordan & Gilbert. (1123)
1385. **Prionotus palmipes** Mitchill. N. (1124)
1386. **Prionotus alatus**[3] Goode & Bean. B.

---

[1] *Peristedion imberbe* Poey.
Only a very few specimens of this fish are known; all in bad condition, having been taken from the stomachs of deep-water fishes at Havana and Pensacola. Barbels very small, scarcely visible—this character distinguishing the species from the others known in America.
(*Peristedion imberbe* Poey, Memorias, II, 389, 1860. *Peristedion micronemus* Poey, Ann. Lyc. Nat. Hist., IX, 321; Jordan, Proc. U. S. Nat. Mus., 1884.)

[2] I am unable to find any positive evidence of the occurrence of the West Indian *Prionotus punctatus* on the coasts of the United States, all the specimens so named being apparently either *P. scitulus* or *P. palmipes*. *Prionotus punctatus* may therefore be omitted.

[3] *Prionotus alatus* Goode & Bean.
Brownish, with about four faint darker cross-bands; vertical fins uniform, the caudal with a black tip and two paler shades before it; dorsal with the usual black spots; pectorals blotched and clouded. Body rather stout, covered with small, rough scales. Maxillary 3 in head; preopercular, opercular, and humeral spines strong, the latter extending farthest back. Palatine teeth few and feeble. Gill-rakers 1+6, besides some rudiments, the longest 3 in eye. Second dorsal spine longest, half head; first spine strongly serrated in front. Caudal subtruncate. Ninth ray of pectoral longest, reaching base of caudal. Pectoral appendages slender. Head 2½; depth 4, D. X—12. A. 11. P. 13+3. Scales 109; 50 tubes in lat. l. Deep water off Charleston, S. C. (*Goode & Bean.*)
(Goode & Bean, Bull. Mus. Comp. Zoöl., XIX, 1883, 210.)

[115] CATALOGUE OF THE FISHES OF NORTH AMERICA.

§ *Prionotus.*

1387. Prionotus ophryas[1] Jordan & Swain. W.
1388. Prionotus stearnsi[2] Jordan & Swain. W.
1389. Prionotus tribulus Cuv. & Val. S. (1125)
1390. Prionotus evolans[3] Linnæus. S. (1126)
1391. Prionotus strigatus[4] Mitchill. N. (1126 b.)
1392. Prionotus stephanophrys Lockington. C. B. (1127)

452.—CEPHALACANTHUS Lacépède. (391)

1393. Cephalacanthus volitans Linnæus. N. S. W. (1128)

Family CXXVII.—LIPARIDÆ. (109.)

453.—MONOMITRA[5] Goode. (392)

1394. Monomitra liparina Goode. B. (1129)

454.—CAREPROCTUS Krōyer. (393)

1395. Careproctus gelatinosus Pallas. A. (1130 b.)
1396. Careproctus reinhardti Krōyer. G. (1130 b.)

455.—LIPARIS Linnæus. (394)

§ *Actinochir* Gill.

1397. Liparis major Walbaum. G. (1131)

§ *Liparis.*

1398. Liparis pulchella Ayres. C. (1132)
1399. Liparis gibba Bean. A (1133)
1400. Liparis tunicata Reinhardt. G. (1135)
1401. Liparis liparis Linnæus. G. N. Eu. (1136)
1401b. *Liparis liparis arctica* Gill. (1134)
1402. Liparis ranula Goode & Bean. N. B. (1137)
1403. Liparis montaguei Donovan. N. Eu. (1138)
1404. Liparis calliodon Pallas. A. (1139)
1405. Liparis cyclopus Günther. A. (1140)

§ *Neoliparis* Steindachner.

1406. Liparis mucosa Ayres. C. B. (1141)

---

[1] *Prionotus ophryas* Jordan & Swain. Proc. U. S. Nat. Mus., 1885. Deep water off Pensacola.

[2] *Prionotus stearnsi* Jordan & Swain, l. c. Deep water off Pensacola, lately discovered by Mr. Silas Stearns.

[3] This species should probably retain the name of *Prionotus evolans*, as adopted in the Synopsis, instead of that of *Prionotus sarritor*, since given it by us (p. 974, Proc. U. S. Nat. Mus., 1882, 615). The type of *Trigla evolans* L., recently examined by Dr. Bean, appears to belong to this species.

[4] *Prionotus strigatus* Cuv. & Val. Described in the Synopsis (p. 736) as *Prionotus evolans lineatus*. Mitchill's name *lineatus*, as stated on page 974, was not given as that of a new species, but through a mistaken identification with the European *Trigla lineata* Bloch.

[5] MONOMITRA Goode.

(Goode, Proc. U. S. Nat. Mus., 1883, 109; type *Amitra liparina* Goode; name a substitute for *Amitra*, preoccupied as *Amitrus*. (Μονος, lacking; μιτρα, stomacher.)

Family CXXVIII.—CYCLOPTERIDÆ. (110)

456.—CYCLOPTERICHTHYS Steindachner. (395)

1407. Cyclopterichthys ventricosus Pallas. A. (1142)
1408. Cyclopterichthys stelleri Pallas. A. (1143)

457.—EUMICROTREMUS Gill. (395 b.)

1409. Eumicrotremus spinosus Müller. A. (1144)

458.—CYCLOPTERUS Linnæus. (396)

1410. Cyclopterus lumpus Linnæus. N. G. Eu. (1145)

Family CXXIX.—GOBIESOCIDÆ. (111)

459.—GOBIESOX Lacépède. (397)

1411. Gobiesox mæandricus Girard. C. (1146)
1412. Gobiesox strumosus Cope. S. (1147)
1413. Gobiesox virgatulus Jordan & Gilbert. S. W. (1147 b.)
1414. Gobiesox rhessodon Rosa Smith. P. (1148)
1415. Gobiesox adustus[1] Jordan & Gilbert. P.
1416. Gobiesox zebra[2] Jordan & Gilbert. P.
1417. Gobiesox erythrops[3] Jordan & Gilbert. P.
1418. Gobiesox eos[4] Jordan & Gilbert. P.

Family CXXX.—BATRACHIDÆ. (112)

460.—BATRACHUS Bloch & Schneider. (398)

1419. Batrachus tau Linnæus. N. S. W. (1149)
1419 b. *Batrachus tau pardus* Goode & Bean. S. (1149 b.)

461.—PORICHTHYS Girard. (399)

1420. Porichthys margaritatus[5] Richardson. C. (1150)
1421. Porichthys porosissimus[6] Cuv. & Val. W. (1150 b.)

---

[1] *Gobiesox adustus* Jordan & Gilbert, Proc. U. S. Nat. Mus., 1881, 360. Mazatlan, southward.

[2] *Gobiesox zebra* Jordan & Gilbert, Proc. U. S. Nat. Mus., 1881, 359. Mazatlan.

[3] *Gobiesox erythrops* Jordan & Gilbert, Proc. U. S. Nat. Mus., 1881, 360. Mazatlan; Tres Marias.

[4] *Gobiesox eos* Jordan & Gilbert, Proc. U. S. Nat. Mus., 1881, 360. Mazatlan.

[5] *Porichthys margaritatus* (Richardson.)

The Pacific species, found from Vancouver's Island to Panama, most abundant northward. The description on page 751 belongs here, and the names *margaritatus* and *notatus*, as also all Pacific coast references to *P. porosissimus*.

[6] *Porichthys porosissimus* (Cuv. & Val.)

The Atlantic species, found from Surinam to Galveston, Pensacola, and Charleston, distinguished from *P. margaritatus* by the strong, unequal palatine teeth, as described on page 958. The names *porosissimus* and *plectrodon* belong to this species, the only one of its genus yet known from the Atlantic.

Family CXXXI.—TRICHODONTIDÆ. (102 b.)

462.—TRICHODON Steller. (337)
1422. Trichodon trichodon Tilesius. A. (975)
1423. Trichodon japonicus [1] Steindachner. A.

Family CXXXII.—LEPTOSCOPIDÆ. (113)

463.—DACTYLOSCOPUS Gill. (400)
1424. Dactyloscopus mundus [2] Gill. P.
1425. Dactyloscopus pectoralis [3] Gill. P.
1426. Dactyloscopus tridigitatus Gill. W. (1151)

464.—MYXODAGNUS [4] Gill.
1427. Myxodagnus opercularis Gill. P.

Family CXXXIII.—URANOSCOPIDÆ. (103)

465.—UPSILONPHORUS [5] Gill. (338)
1428. Upsilonphorus y-græcum Cuv. & Val. S. (976)
1429. Upsilonphorus guttatus Abbott. N. S. (977)

---

[1] *Trichodon japonicus* Steindachner.

Form of body and coloration of *T. trichodon*. First dorsal high, triangular, formed of ten slender spines, and separated by a long interval from the second dorsal. Preopercle with five sharp spines; the two spines on the preorbital very small. Pectoral well developed, all its rays simple, the lower a little thickened; the fin considerably longer than the head and reaching past the last spine of the dorsal. Anal fin with its rays gradually longer posteriorly. Dentition as in *T. trichodon*, the mouth rather more oblique than in the latter. Head 3⅘: depth 3¾. D. X–13; A. 31; P. 25; L. 4¼ inches. Strietok, in the sea of Japan, and Sitka, Alaska (*Steindachner*).

(Steindachner, Ichth., Beitr., X, 4, 1881.)

[2] *Dactylagnus mundus* Gill, Proc. Ac. Nat. Sci. Phila., 1862, 505. Jordan & Gilbert, Proc. U. S. Nat. Mus., 1882, 628. Cape San Lucas to Panama.

We find very small pseudobranchiæ present in living examples of *Dactyloscopus tridigitatus*. Probably none of the family are wholly destitute of these organs.

[3] *Dactyloscopus pectoralis* Gill, Proc. Ac. Nat. Sci. Phila., 1861, 267. Cape San Lucas.

[4] MYXODAGNUS Gill.

(Gill, Proc. Ac. Nat. Sci. Phila., 1861, 269, 270; type *Myxodagnus opercularis* Gill.)

This genus differs from *Dactyloscopus* in the form of the head, which is elongate-conoid, the lower jaw obtusely pointed and provided with a short flap in front. The pseudobranchiæ are well developed and the dorsal fin commences far behind the nape. One species known. (*Myxodes*, a genus of blennies; αγρος, an old name of *Uranoscopus scaber*.) *Myxodagnus opercularis* Gill, l. c., 270. Cape San Lucas.

[5] Instead of genus *Astroscopus* as given in the Synopsis (p. 627) read:

UPSILONPHORUS Gill.

(Gill, Proc. U. S. Nat. Mus., 1861, 113; type *Uranoscopus y-græcum* Cuv. & Val.)

The definition of *Astroscopus* in the text applies entirely to this genus. (Υψιλον, ὑ: φορεω, to bear.)

The species of this genus should stand as:

Upsilonphorus y-græcum (C. & V.) Gill.

The comparison made on page 941 between *A. y-græcum* and *A. anoplus* should be suppressed, as the specimens there called *anoplus* were the young of *y-græcum*, and the differences noted are the changes produced by age.

Upsilonphorus guttatus (Abbott) Gill.

This is the species called *Astroscopus anoplus* by Bean (Proc. U. S. Nat. Mus., 1879, 60) and by us in the text on page 629. The original *anoplus* is, however, very different.

466.—ASTROSCOPUS[1] Brevoort.

1430. **Astroscopus anoplus** Cuv. & Val. S.

Family CXXXIV.—OPISTHOGNATHIDÆ. (103 b.)

467.—GNATHYPOPS Gill. (338 b.)

1431. **Gnathypops rhomaleus**[2] Jordan & Gilbert. P.
1432. **Gnathypops mystacinus**[3] Jordan. W.
1433. **Gnathypops maxillosus** Poey. W.

468.—OPISTHOGNATHUS Cuv. & Val. (339 b.)

1434. **Opisthognathus scaphiura** Goode & Bean. W. (977 c.)
1435. **Opisthognathus lonchura** Jordan & Gilbert. W. (977 d.)
1436. **Opisthognathus punctata**[4] Peters. P.

---

[1] ASTROSCOPUS Brevoort.
(*Agnus* Günther.)

(Brevoort MSS.; Gill, Proc. Ac. Nat. Sci., Phila., 1860, 20; type *Uranoscopus anoplus*. C. & V.)

This genus is distinguished from *Upsilonphorus* chiefly by the armature of the head, which is entirely covered above by a rugose coat of mail as in *Uranoscopus*. In other respects it agrees with *Upsilonphorus*, which should, perhaps, be regarded as a subgeneric section of *Astroscopus*. One species known.

Astroscopus anoplus (Cuv. & Val.).

Jet black above and on lower jaw and spinous dorsal; belly and other fins whitish; top of head with no naked areas except at base of premaxillary; cheeks covered with smooth skin except the narrow suborbital and a long slender preorbital strip lying along the maxillary. A transverse depression behind the eyes; occipital ridges prominent, bluntish. Humeral spine obsolete; preopercle with two blunt processes, the lower turned downwards and forwards. Scales minute, obsolete below; no intralabial filament. Head as broad as deep; head 2½; depth 3¼. D. IV-14; A. 13. New York to Key West. No specimens known more than 2½ inches in length.

Uranoscopus anoplus C. & V., VIII, 493, 1831. *Agnus anoplus* Günther, II, 229 (not *Astroscopus anoplus* of most recent authors).

[2] *Opisthognathus rhomaleus* Jordan & Gilbert, Proc. U. S. Nat. Mus., 1881, 276. Gulf of California.

[3] *Gnathypops mystacinus* Jordan, Proc. U. S. Nat. Mus., 1884.

[4] *Opisthognathus punctatus* Peters, Berliner Monatsberichte, 1869; Jordan, Proc. Ac. Nat. Sci. Phila., 1883, 290. Mazatlan.

Head everywhere finely speckled with black, the body more coarsely and irregularly spotted. Pectoral finely and closely speckled, its edge plain. Ventral fin dusky, similarly marked. Dorsal without large black blotch, finely spotted, the spots behind gradually forming the boundaries of white ocelli, the base of the fins having rings of white around black spots, the upper part with dark rings around pale spots. Caudal with pale spots, its edge, like that of the dorsal, somewhat dusky, not black. Anal with a broad, blackish edge, and with dark spots, those near the base of the fin largest. Lining membrane of maxillary with the usual bands of white and inky black.

Scales very small, about 125 in lateral line. Dorsal spines continuous with the soft rays. D. 28; A. 18. No vomerine teeth. Maxillary very long, extending slightly beyond head.

Only the type of this species is yet known.

Family CXXXV.—CHIASMODONTIDÆ. (120 b.)

469.—CHIASMODON Johnson. (446)

1437. **Chiasmodon niger** Johnson. B. (1250)

Family CXXXVI.—BLENNIIDÆ. (114)

470.—OPHIOBLENNIUS Gill. (401)

1438. **Ophioblennius webbi** Valenciennes. W. P. (1152)

471.—CHASMODES Cuv. & Val. (402)

1439. **Chasmodes bosquianus** Lacépède. S. (1153)
1440. **Chasmodes quadrifasciatus** Wood. S. (1154)
1441. **Chasmodes saburræ** Jordan & Gilbert. S. (1154 b.)

472.—HYPSOBLENNIUS[1] Gill. (403)

1442. **Hypsoblennius brevipinnis**[2] Günther. P.
1443. **Hypsoblennius gentilis** Girard. C. P. (1155 b.)
1444. **Hypsoblennius gilberti** Jordan. C. (1155)
1445. **Hypsoblennius punctatus**[3] Wood. S. (1156, 1156 b.)
1446. **Hypsoblennius ionthas** Jordan & Gilbert. S. (1156 c.)
1447. **Hypsoblennius scrutator** Jordan & Gilbert. S. (1156 d.)

473.—HYPLEUROCHILUS Gill. (404)

1448. **Hypleurochilus multifilis** Girard. S. (1157)
1449. **Hypleurochilus geminatus** Wood. S. (1158)

474.—BLENNIUS Linnæus. (405)

§ *Blennius.*

1450. **Blennius stearnsi**[4] Jordan & Gilbert. W. (1159 b.)
1451. **Blennius favosus** Goode & Bean. W. (1159 c.)
1452. **Blennius asterias** Goode & Bean. W. (1159 d.)

§ *Pholis* Cuv. & Val.

1453. **Blennius carolinus** Cuv. & Val. S. (1160)

---

[1] The generic name *Hypsoblennius* Gill (Cat. Fish. East Coast U. S., 1861; *H. hentzi*) introduced without definition or explanation is equivalent to *Isesthes* Jordan & Gilbert. If it be thought best to adopt such *nomina nuda*, *Hypsoblennius* has precedence over *Isesthes*.

[2] *Blennius brevipinnis* Günther, Cat. Fishes, III, 226. Mazatlan, southward. This species is a genuine *Isesthes*, as is also the *Blennius striatus* of Steindachner, from Panama.

[3] *Isesthes hentzi* should be erased. It is identical with *Isesthes punctatus*, as given on page 758 of the Synopsis.

[4] *Blennius fucorum* should be erased. It is a tropical species introduced into our faunal lists by DeKay, on information which was probably erroneous.

475.—RUPISCARTES Swainson.[1]

1454. **Rupiscartes chiostictus**[2] Jordan & Gilbert. P.
1455. **Rupiscartes atlanticus**[3] Cuv. & Val. P. W.

476.—EMBLEMARIA [4] Jordan & Gilbert.

1456. **Emblemaria nivipes** Jordan & Gilbert. W. P.

477.—NEOCLINUS Girard. (406)

1457. **Neoclinus satiricus** Girard. C. (406)
1458. **Neoclinus blanchardi** Girard. C. (1162)

478.—LABROSOMUS Swainson.

1459. **Labrosomus nuchipinnis** Quoy & Gaimard. W. (1163)
1459 b. *Labrosomus nuchipinnis xanti*[5] Gill. P.
1460. **Labrosomus zonifer**[6] Jordan & Gilbert. P.

---

[1] RUPISCARTES Swainson.

(Swainson, Class'n Anim., 1839, II, 275; type *Salarias alticus* C. & V.)
 As here understood, this genus differs from *Blennius*, in having the teeth in the jaws slender and movable. From the genus *Salarias* Cuv. (type *S. quadripinnis* Cuv.), which has the same dentition, and to which genus its species have been usually referred, it differs in the presence of posterior canines. Species numerous, in tide pools of the tropics. (Latin, *rupis*, rock; σκάρτης, a leaper; "it is said to jump on the sea-rocks like a lizard"; *Swainson*.)
 [2] *Salarias chiostictus* Jordan & Gilbert, Proc. U. S. Nat. Mus., 1881, 363. Mazatlan.
 [3] *Salarias atlanticus* Cuv. & Val., XI, 321; Günther, III, 242. Tropical America, on both coasts, north to Cape San Lucas.

[4] EMBLEMARIA Jordan & Gilbert.

(Jordan & Gilbert, Proc. U. S. Nat. Mus., 1882, 627; type *Emblemaria nivipes* Jordan & Gilbert.)
 Body moderately elongate, not compressed, naked. Ventrals jugular, I, 2. Dorsal fin continuous, beginning at the nape, not confluent with the caudal. Spines and soft rays similar, both much elevated. Head cuboid, formed much as in *Opisthognathus*. Lower jaw very acute at symphysis. A single series of strong, blunt, conical teeth on each jaw and on vomer and palatines. Teeth of vomer and palatines larger, forming a uniform curve. No cirri. Gill openings very wide, the membranes broadly united below, free from the isthmus. One species known. (*Emblema*, a banner (emblem); from the elevated fins.)
 Emblemaria nivipes Jordan & Gilbert, Proc. U. S. Nat. Mus., 1882, 627.
 Originally described from the Pearl Islands (Panama). A specimen which we cannot distinguish from this species was obtained at Pensacola by Mr. Silas Stearns. See Proc. U. S. Nat. Mus., 1884.
 [5] *Labrosomus xanti* Gill, Proc. Ac. Nat. Sci. Phila., 1860, 107; *Clinus xanti* Jordan & Gilbert, Proc. U. S. Nat. Mus., 1882, 368. Gulf of California, southward. The genus *Labrosomus*, as here understood, differs from *Clinus* chiefly in the absence of the upturned spine-like process on the inner edge of the shoulder girdle, characteristic of the latter genus and *Heterostichus*. This process is found on *Clinus acuminatus*, the type of the genus *Clinus*.
 [6] *Clinus zonifer* Jordan & Gilbert, Proc. U. S. Nat. Mus., 1881, 361. Mazatlan.

479.—TRIPTERYGION [1] Risso.

1461. Tripterygion carminale [2] Jordan & Gilbert. P.

480.—CLINUS Cuv. & Val. (407)

§ *Gibbonsia* Cooper.

1462. Clinus evides Jordan & Gilbert. C. (1164)

481.—HETEROSTICHUS Girard. (408)

1463. Heterostichus rostratus Girard. C. (1165)

482.—CREMNOBATES Günther. (409)

1464. Cremnobates altivelis [3] Lockington. P.
1465. Cremnobates marmoratus Steindachner. W. (1166b.)
1466. Cremnobates fasciatus [4] Steindachner. W.
1467. Cremnobates affinis [5] Steindachner. W.

---

[1] TRIPTERYGION Risso.

(Risso, Europe Méridion. 1826, III, 241; type *Blennius tripteronotus* Risso.) This genus is allied to *Clinus*, differing chiefly in the division of the dorsal fin into three nearly or quite separate fins, the anterior of 3 to 6 spines, the median one of many spines and the last of many soft rays. Warm seas in tide-pools. ($T\rho\epsilon\tilde{\iota}\varsigma$, three; $\pi\tau\epsilon\rho\acute{\upsilon}\gamma\iota o\nu$, fin.)

[2] *Tripterygium carminale* Jordan & Gilbert, Proc. U. Nat. Mus., 1881, 362. Mazatlan to Panama.

[3] *Cremnobates altivelis* Lockington, Proc. Ac. Nat. Sci. Phila., 1881. Gulf of California.

[4] *Cremnobates fasciatus* Steindachner.

Light pinkish-brown, much mottled, and with 6 or 8 darker bars; sides of head marbled with whitish, its cirri pale; 3 black spots behind and below eye; dorsal pale, with 9 blackish blotches extending from the bands on the sides; in the next the last of these is a large blue-black spot ocellated with orange; anal with 5 dark blotches and no ocellus; a dark band across base of caudal; caudal otherwise pale yellowish with dark dots. Pectorals whitish, barred with black; its base with a whitish area; with a brown center, below which is a small black spot. Ventrals barred. Body rather slender, a little deeper than as in *C. integripinnis*, the snout less acute than in *C. marmoratus*. First dorsal spine rather higher than second, and lower than the spines of posterior part of fin; membrane of third spine joining second dorsal at a point above its base, the two parts of the fin therefore separated only by an emargination. Tentacle above eye slender, small; cirri on side of occiput bluish. Head 4; depth 4¼. D. III, 24, 1. A. II, 18. Lat. l. 37. L. 2 inches. Florida Straits; north to Key West.

(Steindachner, Ichth. Beitr, V, 1876, 176). For a comparison of our species of *Cremnobates*, see Jordan, Proc. U. S. Nat. Mus., 1884, 142.)

[5] *Cremnobates affinis* Steindachner.

Dark brown, paler than in *C. nox*, but darker and more uniform than in *C. fasciatus*; lower side of head pearly gray, thickly speckled with darker; sides with 5 very faint darker cross-bands; dorsal and anal dusky, the latter with a pale edge; between the 18th and 22d dorsal spines a large dark spot ocellated with yellowish; caudal yellowish white, with darker cross-streaks; a blackish band at its base; pectoral dusky at base, its posterior half yellowish, with darker cross-streaks; ventral similar. A wedge-shaped whitish band extending backward from eye to opercle. Form of *C. integripinnis*; maxillary reaching to below posterior margin of eye; a fringed tentacle above eye and one on each side of occiput. First dorsal low, its longest (second) ray

1468. Cremnobates integripinnis Rosa Smith. C. P. (1166)
1469. Cremnobates nox[1] Jordan & Gilbert. W.

483.—CHIROLOPHUS Swainson. (410)

1470. Chirolophus polyactocephalus[2] Pallas. A. (1167)

484.—MURÆNOIDES[3] Lacépède. (411)

1471. Murænoides gunnellus Linnæus. N. G. Eu. (1168)
1472. Murænoides fasciatus Bloch & Schneider. G. (1169)
1473. Murænoides ornatus Girard. A. (1170)
1474. Murænoides maxillaris Bean. A. (1171)
1475. Murænoides dolichogaster Pallas. H. (1172)

485.—APODICHTHYS Girard. (412)

1476. Apodichthys flavidus Girard. C. (1174)
1477. Apodichthys fucorum Jordan & Gilbert. C. (1175)
1478. Apodichthys univittatus[4] Lockington. P.

486.—ANOPLARCHUS Gill. (413)

1479. Anoplarchus atropurpureus[5] Kittlitz. C. A. (1176)

487.—XIPHISTER Jordan. (414)

1480. Xiphister chirus Jordan & Gilbert. C. (1178)
1481. Xiphister mucosus[6] Girard. C. (1179)
1482. Xiphister rupestris Jordan & Gilbert. C. (1180)

488.—CEBEDICHTHYS Ayres. (415)

1483. Cebedichthys violaceus Girard. C. (1181)

489.—EUMESOGRAMMUS Gill. (416)

1484. Eumesogrammus præcisus Kröyer. G. (1182)
1485. Eumesogrammus subbifurcatus Storer. N. (1183)

490.—STICHÆUS Reinhardt. (417)

1486. Stichæus punctatus Fabricius. G. (1184)

---

shorter than the highest of second dorsal; membrane of third spine joining the fourth spine just above its base. Last ray of second dorsal joined by membrane to base of caudal. Head 4; depth 4⅔, D. III, 27, I. A. II, 19. V. 1, 2. Lat. 1. 33 to 35. Key West; St. Thomas.

(Steindachner, Ichthyologische Beiträge, V, 178, 1876. Jordan, l. c., 142.)

[1] *Cremnobates nox* Jordan & Gilbert, Proc. U. S. Nat. Mus., 1884, 30. Key West.

[2] *Blennius polyactocephalus* Pallas, lately rediscovered by Mr. Nelson in Alaska, proves to be, as supposed in the Synopsis, a genuine species of *Chirolophus*.

[3] I here omit *Murænoides (Asternopteryx) gunelliformis*. It is not certain that the single known specimen is a *Murænoides* or that it is from American waters.

[4] *Apodichthys univittatus* Lockington, Proc. Ac. Nat. Sci. Phila., 1881, 118. Gulf of California.

[5] *Anoplarchus alectrolophus* should not have been inserted. It is an Asiatic species, not found within our limits.

[6] The type of *Xiphidium cruoreum* Cope, examined by Mr. Meek, is identical with *X. mucosus*.

491.—NOTOGRAMMUS Bean. (418)

1487. Notogrammus rothrocki Bean. A. (1185)

492.—LEPTOCLINUS Gill.

1488. Leptoclinus maculatus Fries. G. (1186)

493.—LUMPENUS Reinhardt. (419)

1489. Lumpenus medius Reinhardt. G. (1187)
1490. Lumpenus anguillaris Pallas. A. (1188)
1491. Lumpenus lumpenus Müller. G. (1189)

494.—LEPTOBLENNIUS Gill. (420)

1492. Leptoblennius nubilus Richardson. G. (1190)
1493. Leptoblennius serpentinus Storer. N. (1191)
1494. Leptoblennius lampetræformis Walbaum. G. (1192)

495.—PHOLIDICHTHYS [1] Bleeker.

1495. Pholidichthys anguilliformis Lockington. P.

Family CXXXVII.—CRYPTACANTHODIDÆ.[2]

496.—DELOLEPIS Bean. (421)

1496. Delolepis virgatus Bean. A. (1193)

497.—CRYPTACANTHODES Storer. (422)

1497. Cryptacanthodes maculatus Storer. N. (1194)

Family CXXXVIII.—ANARRHICHADIDÆ.[2]

498.—ANARRHICHAS Linnæus. (423)

1498. Anarrhichas lupus Linnæus. N. Eu. (1195)
1499. Anarrhichas minor Olafsen. G. Eu. (1196)
1500. Anarrhichas latifrons Steenstrup & Halgrimsson. G. Eu. (1197)
1501. Anarrhichas lepturus Bean. A. (1198)

499.—ANARRHICHTHYS Ayres. (424)

1502. Anarrhichthys ocellatus Ayres. C. (1199)

---

[1] PHOLIDICHTHYS Bleeker.

(Bleeker, Boeroe, 406; type *Pholidichthys leucotænia* Bleeker.)
Body elongate, tapering, naked; snout obtuse; no cirri. Teeth unequal, on jaws only. Dorsal, anal, and caudal fins distinct, but connected by membrane, the dorsal formed of flexible spines. Ventrals inserted scarcely before the pectorals, of two rays. Two species known, of the tropical parts of the Pacific. ($\Phi o \lambda i \varsigma$, Pholis; $i\chi\theta\dot{\upsilon}\varsigma$, fish.)

Pholidichthys anguilliformis Lockington, Proc. Ac. Nat. Sci. Phila., 1881, 118. Dredged off Amortiguado Bay, Gulf of California.

[2] There seems to be no doubt that the families of *Cryptacanthodidæ* and *Anarrhichadidæ* at least, should be detached from the *Blenniidæ*. Whether the latter group should be further subdivided or not, I am not certain. In the northern types (*Xiphisterinæ*, *Stichæinæ*) the vertebræ are much more numerous than in the tropical *Clininæ* and *Blenniinæ*.

Family CXXXIX.—LYCODIDÆ. (115)

500.—ZOARCES Cuvier. (425)

1503. Zoarces anguillaris Peck. N. G. (1200)

501.—LYCODOPSIS Collett. (426)

1504. Lycodopsis pacificus Collett. C. A. (1201)
1505. Lycodopsis paucidens Lockington. C. (1202)

502.—LYCODONUS[1] Goode & Bean.

1506. Lycodonus mirabilis Goode & Bean. B.

503.—LYCENCHELYS[2] Gill.

1507. Lycenchelys paxillus Goode & Bean. B. (1203)
1508. Lycenchelys paxilloides[3] Goode & Bean. B.
1509. Lycenchelys verrilli Goode & Bean. B.

504.—LYCODES Reinhardt. (427)

1510. Lycodes vahli Reinhardt. B. G. (1205)
1511. Lycodes esmarki Collett. B. G. Eu. (1206)
1512. Lycodes reticulatus Reinhardt. B. G. (1207)
1513. Lycodes seminudus Reinhardt. B. G. (1208)
1514. Lycodes nebulosus Reinhardt. G. (1209)
1515. Lycodes coccineus Bean. A. (1210)

---

[1] LYCODONUS Goode & Bean.

(Goode & Bean, Bull. Mus. Comp., Zoöl., XIX, 1883, 208; type *Lycodonus mirabilis* Goode & Bean.)

Body elongate, formed as in *Lycenchelys*. Scales small, circular, imbedded in the skin; lateral line very short, obsolete posteriorly. Jaws without fringes, lower jaw included. Fin rays all articulated, each ray of dorsal and anal supported laterally by a pair of sculptured scutes. Caudal distinct, not fully connate with dorsal and anal. Ventrals present. Gill opening narrow. Teeth as in *Lycodes*. Deep water (*Lycodes*; *Onos*).

Lycodonus mirabilis Goode & Bean.

Form of *Lycenchelys verrilli*, very slender; head, nape, and fins scaleless; maxillary reaching front of pupil. Dorsal inserted slightly behind base of pectorals. Length of pectorals 3 times snout. Eye $2\frac{1}{4}$ in head, $3\frac{1}{4}$ times interorbital width. Head 7; depth 18. D. 80 +. A. 70 +. Gulf Stream, lat. 40°.
(Goode & Bean, Bull. Mus. Comp. Zoöl., XIX, 1883, 208.)

[2] LYCENCHELYS Gill.

(Gill, Proc. Ac. Nat. Sci., Phila., 1884, 180; type *Lycodes muræna* Collett.)

This name *Lycenchelys* may be used for Collett's second group, which have the body elongate; height of the body contained from 12 to 24 times in the total length (*Gill*). (Λύκος, wolf; ἔγχελυς, eel.)

[3] *Lycenchelys paxilloides* Goode & Bean.

Light brown, the head somewhat darker. Form of *L. paxillus*, but with a smaller mouth and less prominent cheeks. Dorsal beginning over tip of pectoral; ventral little longer than pupil. Scales very small, present everywhere except on head and pectorals, nearly covering vertical fins. Eye $3\frac{1}{4}$ in head, equal to snout, which is 4 times interorbital width. Head 8, depth 16. D. (with half caudal) 118. A. 110. P. 16. V. 3. Gulf Stream, lat. 40°, in deep water (*Goode & Bean*).

(*Lycodes paxillus* Goode & Bean. Bull. Mus. Comp. Zoöl., XIX, 1883, 207.)

505.—LYCODALEPIS Bleeker. (428)

1516. **Lycodalepis mucosus** Richardson. G. (1211)
1517. **Lycodalepis turneri** Bean. A. (1212)
1518. **Lycodalepis polaris** Sabine. G. (1213)

506.—GYMNELIS Reinhardt. (429)

1519. **Gymnelis viridis**[1] Fabricius. G. A. (1214, 1215?)

507.—LYCOCARA[2] Gill. (430)

1520. **Lycocara parrii** Ross. G. (1216)

508.—MELANOSTIGMA[3] Günther.

1521. **Melanostigma gelatinosum** Günther. B.

Family CXL.—CERDALIDÆ.[4]

509.—MICRODESMUS.[5] Günther.

1522. **Microdesmus dipus** Günther. P.

---

[1] I here omit *Gymnelis stigma*. It is probably based on an inaccurate description of *Gymnelis viridis*. If, however, really possessing scales, it may belong to the Antarctic genus *Maynea* (Cunningham), which differs from *Lycodes* chiefly in the absence of ventrals.

[2] LYCOCARA Gill.

(Gill, Proc. Ac. Nat. Sci. Phila., 1884, 180; type *Ophidium parrii* Ross.)
This name is a substitute for *Uronectes*, which is preoccupied. ($\Lambda\nu\kappa o\varsigma$, wolf; $\kappa\alpha\rho\alpha$, head.)

[3] MELANOSTIGMA Günther.

(Günther, Proc. Zoöl. Soc. Lond., 1881, 21; type *Melanostigma gelatinosum* Günther.)
Allied to *Gymnelis*; "technically distinguished by the much more elongate teeth, which in the jaws, as well as on the vomer and palatines, stand in single series." Gill openings much smaller than in related forms, reduced to a small foramen above the base of the pectoral. Skin loose and movable, as in *Liparis*, enveloping the vertical fins; pectorals very small; ventrals, none. Body tapering very rapidly backward; the tail very slender. Deep sea. ($M\epsilon\lambda\alpha\varsigma$, black; $\sigma\tau\iota\gamma\mu\alpha$, spot.)

Melanostigma gelatinosum Günther.

Purplish above; sides grayish, marbled with darker, the end of the tail almost black. Head large, deep, compressed; the snout blunt. Eye large, $3\frac{1}{4}$ in head, longer than snout. Cleft of mouth oblique, the maxillary reaching a little past front of pupil, the lower jaw not projecting. Inside of mouth, gill openings and vent black. Dorsal beginning above middle of pectoral, low in front, becoming higher than the part of the body below it posteriorly. Head $6\frac{1}{4}$. Deep waters of the Atlantic; Martha's Vineyard; Straits of Magellan.
(Günther, Proc. Zoöl. Soc. London, 1881, 21; Goode & Bean, Bull. Comp. Zoöl., XIX, 1883, 209.)

[4] I suggest the provisional name *Cerdalidæ* for two closely related genera, *Cerdale* Jordan & Gilbert, and *Microdesmus* Günther, which seem to be allied to the *Lycodidæ*, differing in the small, slit-like gill openings and in the non-isocercal tail. The three known species are scantily represented in collections, and until their osteology is examined we cannot be sure as to their relation to the *Lycodidæ*, *Congrogadidæ*, and *Brotulidæ*.

[5] MICRODESMUS Günther.

Günther, Proc. Zool. Soc., London, 1864, 26; type *Microdesmus dipus* Günther.)
Body anguilliform, covered with rudimentary scales. Head small, with short snout and small mouth; lower jaw projecting. Teeth minute, in jaws only. Gill opening reduced to a very narrow, somewhat oblique slit, in front of lower part of pectorals. Vertical fins well developed, the dorsal and anal joined to the caudal by a thin mem-

Family CXLI.—CONGROGADIDÆ. (116)

510.—SCYTALISCUS[1] Jordan & Gilbert. (431)

1523. Scytaliscus cerdale Jordan & Gilbert. A. (1217)

Family CXLII.—FIERASFERIDÆ. (117)

511.—FIERASFER Cuvier. (432)

1524. Fierasfer dubius[2] Putnam. P. W. (1218)

Family CXLIII.—OPHIDIIDÆ. (118)

512.—OPHIDION Linnæus. (433)

1525. Ophidion marginatum[3] Dekay. S. W. (1219, 1220)
1526. Ophidion holbrooki Putnam. W. (1221)
1527. Ophidion beani[4] Jordan. W. (1221 b.)

513.—OTOPHIDIUM[5] Gill. (433 b.)

1528. Otophidium taylori Girard. C. (1222)
1529. Otophidium omostigma Jordan & Gilbert. W. (1223 b.)

514.—LEPTOPHIDIUM Gill.

1530. Leptophidium profundorum Gill. W. B. (1223)

Family CXLIV.—BROTULIDÆ.[6] (119)

515.—BYTHITES Reinhardt. (434)

1531.—Bythites fuscus Reinhardt. G. (1224)

---

brane. Tail not isocercal. Rays of dorsal all articulate; all but a few of the last simple. Ventral fins very small, reduced to a single ray. Pectorals moderate. Vent normal. Pacific coast of tropical America. ($M\iota\kappa\rho o\varsigma$, small; $\delta\varepsilon\delta\mu o\varsigma$, a band.)

Microdesmus dipus Günther, l. c.

Gulf of California to Panama. The two remaining species of this family, *Microdesmus retropinnis* and *Cerdale ionthas*, both from Panama, are described by Jordan & Gilbert, Bull. U. S. Fish Comm., 1881, 331.

[1] SCYTALISCUS Jordan & Gilbert.

Proc. U. S. Nat. Mus., 1883, 111; name a substitute for *Scytalina*, preoccupied in Coleoptera as *Scytalina* Erichson. It is doubtful whether this genus is really an ally of *Congrogadus*.

[2] *Fierasfer dubius* Putnam = *Fierasfer arenicola* Jordan & Gilbert, Proc. U. S. Nat. Mus., 1881, 363. Mazatlan. See Jordan & Gilbert, Proc. U. S. Nat. Mus., 1882, 629.

Ophidium josephi Girard and *Ophidium graëllsi* Poey (not of Jor. & Gilb.) seem to be identical with *O. marginatum*.

[3] The species described in the Synopsis as *Ophidium graëllsi* should stand as *Ophidion beani* Jordan & Gilbert. See Proc. U. S. Nat. Mus., 1883, 143.

[4] OTOPHIDIUM Gill, gen. nov.

Type *Genypterus omostigma* Jordan & Gilbert. This genus differs from *Ophidium* in the presence of a sharp concealed spine on the opercle. The typical species has been wrongly referred to *Genypterus*.

[5] The Brotuline genera (*Bythites* and *Dinematichthys*) have been erroneously placed in the Synopsis among the *Gadidæ*. For the characters of the *Brotulidæ* see Gill, Proc. Ac. Nat. Sci. Phila., 1863, 252; 1864, 200, and 1884, 169, 175. These fishes are certainly much nearer the *Ophidiidæ*, or even the *Lycadidæ*, than the *Gadidæ*.

516.—DINEMATICHTHYS Bleeker. (435)

§ *Brosmophycis* Gill.

1532. Dinematichthys marginatus Ayres. C. (1225)
1533. Dinematichthys ventralis[1] Gill. P.

517.—BARATHRODEMUS[2] Goode & Bean.

1534. Barathrodemus manatinus Goode & Bean. B.

518.—DICROLENE[3] Goode & Bean. B.

1535. Dicrolene intronigra Goode & Bean. B.

---

[1] *Brosmophycis ventralis* Gill, Proc. Ac. Nat. Sci. Phila., 1863, 253. Cape San Lucas, southward.

[2] BARATHRODEMUS Goode & Bean.
(Goode & Bean, Bull. Mus. Comp. Zoöl., XIX, 1883, 200; type *Barathrodemus manatinus* G. & B.)
Body brotuliform, much compressed; head compressed; mouth moderate. Head unarmed, except for a short flattened spine at upper angle of opercle. Snout long, projecting far beyond premaxillaries, its tip much swollen; jaws subequal in front. Teeth minute, in villiform bands on jaws, vomer and palatines. No barbels. Anterior nostrils on the outer angles of the dilated snout, circular, each surrounded by a cluster of mucous tubes. Posterior nostrils above front of eye. Gill openings wide, the membranes not united. Gill-rakers rather few. Body and head covered with small, thin, scarcely imbricated scales. Dorsal and anal long. Caudal fin separate, long, and slender. Ventrals close together, far in front of pectorals, each reduced to a single bifid ray. Deep-sea fishes. ($\beta\acute{\alpha}\rho\alpha\theta\rho o\nu$, a gulf or deep abyss; $\delta\tilde{\eta}\mu o\varsigma$, people.)

Barathrodemus manatinus Goode & Bean.
Grayish brown; abdomen black. Snout longer than eye, its form resembling that of the manatee. Maxillary reaching to opposite front of eye, its length 2¼ in head. Eye 5¼ in head. Insertion of dorsal above that of pectoral. Ventrals inserted nearly below middle of opercle, their length half head. Head 6; depth 7¼. D. 106; A. 86; C + 5 +; Lat. l. 175. Gulf Stream, latitude 33°. (*Goode & Bean.*)
(Goode & Bean, Bull. Mus. Comp. Zoöl., XIX, 1883, 200.)

[3] DICROLENE Goode & Bean.
(Goode & Bean, Bull. Mus. Comp. Zoöl., 1883, 202, XIX; type *Dicrolene introniger* G. & B.)
Body brotuliform, moderately compressed; head somewhat compressed, the mouth large; tip of maxillary much dilated. Eye large, placed high. Head with supraorbital spines; several strong spines on the preopercle and one long spine at upper angle of opercle. Snout short, not projecting; jaws subequal. Teeth in narrow, villiform bands on jaws, head of vomer, and on palatines. No barbel. Gill membranes separate. Caudal fin small, separate. Dorsal and anal fins long. Pectoral with several of its lower rays separate and very much produced. Ventrals close together, under front of operculum, each composed of a single bifid ray. Head and body covered with small scales. Lateral line incomplete. Stomach siphonal; pyloric cæca rudimentary; intestine short. Deep water. ($\Delta\acute{\iota}\chi\rho oo\varsigma$, forked; $\dot{\omega}\lambda\acute{\epsilon}\nu\eta$, arm.)

Dicrolene introniger Goode & Bean.
Opercular spine with its exposed portion half as long as eye, which is as wide as interorbital space, and 4 in head. Mouth large, the maxillary extending beyond eye, its length considerably more than half head; width of expanded tip of maxillary ⅔ eye. Bones of head with large muciferous cavities. Length of caudal half distance from

519.—BASSOZETUS Gill.[1]

1536. **Bassozetus normalis** Gill. B.

Family CXLV.—GADIDÆ. (120)

520.—RHINONEMUS Gill.

1537. **Rhinonemus cimbrius** Linnæus. N. Eu. (1226)

521.—ONOS[2] Risso. (436)

1538. **Onos reinhardti** Kröyer. G. (1227)
1539. **Onos ensis** Reinhardt. G. (1228)
1540. **Onos rufus**[3] Gill. B.
1541. **Onos septentrionalis**[4] Collett. G. Eu.

---

snout to front of dorsal. Eight lower rays of pectorals free, much prolonged, the longest and most anterior being nearly one-third length of body and more than three times length of the nearest of the normal rays, which are, however, about equal to the least of the free rays; normal rays of pectorals 4 in body. Head 5; depth 6. D. 100; A. ca. 85, C. 7; P. 19 + 7; Lat. l. ca. 115. Gulf Stream, latitude 34°. (*Goode & Bean.*)

(Goode & Bean, l. c. 202.)

[1] BASSOZETUS Gill.

(Gill, Proc. U. S. Nat., Mus., 1883, 259; type *Bassozetus normalis* Gill.)

"Dinematichthyine brotulids with a slender body; a narrow differentiated caudal fin; anus about a third of the total length from the snout; small eyes, and unarmed head and shoulders." Deep sea. (βάσσων, deep; ϛήτων, seeker.)

Bassozetus normalis Gill. Deep water; latitude 39°.

(Gill, l. c. 259.)

The descriptions, generic and specific in this paper, "Diagnoses of new Genera and Species of Deep-sea Fish-like vertebrates," are among the most brief and unsatisfactory in our ichthyological literature. This paper, by a most able and competent ichthyologist, from the brief and superficial character of its descriptions, is likely to cause great confusion in the study of the Bassalian fauna of the Atlantic, unless soon followed by accurate and sufficient descriptions.

[2] "The *Lotinæ*, and apparently the *Oninæ*, have doubled or paired frontals. * * * It seems probable that they may be segregated in a peculiar family." Gill, Proc. Ac. Nat. Sci. Phila., 1884, 172.

[3] *Onos rufus* Gill.

Color in life almost uniform salmon or brick-red; barbels three; enlarged dorsal ray not shorter than head; some enlarged brown-colored teeth developed in the exterior row. Closely allied to *O. ensis*, but apparently different in color. Deep sea, latitude 40°. (*Gill.*)

(Gill, Proc. U. S. Nat. Mus., 1883, 259.)

[4] *Onos septentrionalis* Collett.

Three barbels, two at the nostrils, one at the chin, besides a row of about eight shorter rudimentary barbels along the edge of the upper lip; eye small, half length of snout; cleft of mouth extending far beyond eye, its length nearly equal to that of postorbital part of head; teeth rather small, unequal; outer teeth of upper jaw and some of the inner teeth of lower enlarged; first ray of first dorsal short, about as long as snout; vent midway between tip of snout and last anal ray; lateral line with about 20 large pores, grayish brown, paler below; cavity of mouth white. D. 50; A. 42; P. 16. Coast of Norway; one specimen known from Greenland. (*Collett.*)

(*Motella septentrionalis* Collett, Ann. Mag. Nat. Hist., 15, 82, 1874; *Onos septentrionalis* Collett, Norske Nord-Havs Exped., 1880, 139.)

522.—LOTA Cuvier. (441)

1542. Lota lota maculosa Le Sueur. Vv. Eu. (1236)

523.—PHYCIS Bloch & Schneider. (437)

1543. Phycis regius Walbaum. N. S. (1229)
1544. Phycis floridanus[1] Bean & Dresel. S.
1545. Phycis earlli Bean. S. (1230)
1546. Phycis chuss Walbaum. N. (1231)
1547. Phycis tenuis Mitchill. N. (1232)
1548. Phycis chesteri Goode & Bean. B. (1233)

524.—LÆMONEMA[2] Günther.

1549. Læmonema barbatula Goode & Bean. B.

525.—ANTIMORA[3] Günther. (438)

1550. Antimora viola Goode & Bean. B. (1233 b.)

---

[1] PHYCIS FLORIDANUS Bean & Dresel.

In general appearance it resembles *P. regius*, differing from this in its smaller scales and more numerous dorsal rays. The greatest height is one-fifth of the total length to caudal base, and equals four-fifths of the length of head. Head 4 times in length to caudal base; eye slightly less than snout, 5 times in length of head; maxilla slightly less than mandible, one-half length of head. First dorsal not produced; ventral about five-fourths length of head; pectoral equal to head in length. Dorsal 13, 57; Anal, 49. Scales between first dorsal and lateral line in nine or ten rows; about 120 scales in the lateral line; L. 7¼ inches. Pensacola. (*Bean & Dresel.*)
(Bean & Dresel, Proc. Biol. Soc. Wash., 1884, 100.)

[2] LÆMONEMA Günther.

(Günther, IV, 356, 1862; type *Phycis yarrelli* Lowe.)
This genus is scarcely distinct from *Phycis*, differing chiefly in the character of the first dorsal, which is composed of five rays only, the anterior ray being filamentous. Deep water. ($\Lambda\alpha\iota\mu o\varsigma$, throat; $\nu\tilde{\eta}\mu\alpha$, thread.)
Læmonema barbatula Goode & Bean.
Color of species of *Phycis*; dorsal and anal with narrow black margins. Eye 3 in head; upper jaw a little more than 2; barbel half as long as eye; vent under 6th ray of spinous dorsal; first ray of first dorsal elongate, about 3 times length of caudal, about reaching 24th ray of second dorsal. Distance from snout to front of anal twice length of head; ventrals as long as pectorals, not reaching vent; scales small, very thin, deciduous. D. 5-63. A. 59. P. 19. V. 2. Scales 13-140, 31. L. 7 inches. Gulf Stream, latitude 32°, in deep water. (*Goode & Bean.*)
(*Læmonema barbatula* Goode & Bean, Bull. Mus. Comp. Zoöl., XIX, 204.)

[3] *Haloporphyrus viola* belongs to the subgenus *Antimora* (Günther, Ann. Mag. Nat. Hist., 1878, 2; type *Haloporphyrus rostratus* Günther). This group differs from *Haloporphyrus* "in the form of the snout, the backward position of the vent, the imperfect division of the anal, in which latter respect it approaches *Mora*." In *Haloporphyrus* the snout is subconical, obtusely rounded; in *Antimora* it forms a flat, triangular lamina, sharply keeled at the sides, resembling the snout of *Macrurus*. The diagnosis of *Haloporphyrus* given in the Synopsis (p. 800) applies to *Antimora* and not to *Haloporphyrus*.
In the very brief description of *Haloporphyrus rostratus* Günther, l. c. (from the mid-Atlantic east of Rio de la Plata), there is nothing by which our species can be distinguished from it. It is probable that the two will prove identical. *A. rostrata* has five months' priority in date over *A. viola*.

526.—PHYSICULUS[1] Kaup. (439)

1551. Physiculus fulvus Bean. B.

527.—LOTELLA[2] Kaup.

1552. Lotella maxillaris Bean. B.

528.—MOLVA Nilsson. (440)

1553. Molva molva Linnæus. G. Eu. (1235)

529.—BROSMIUS Cuvier.

1554. Brosmius brosme Müller. N. G. Eu. (1237)

530.—MELANOGRAMMUS[3] Gill.

1555. Melanogrammus æglefinus Linnæus. N. G. Eu. (1238)

531.—GADUS Linnæus. (443)

1556. Gadus callarias Linnæus. N. G. A. Eu. (1239)
1557. Gadus ogac[4] Richardson. G.

532.—PLEUROGADUS[5] Bean.

1558. Pleurogadus navaga Kölreuter. A. (1240)

533.—MICROGADUS Gill.

1559. Microgadus proximus Girard. C. (1241)
1560. Microgadus tomcod Walbaum. N. (1242)

534.—POLLACHIUS Nilsson.

§ *Pollachius.*

1561. Pollachius virens Linnæus. N. Eu. (1243)
1562. Pollachius chalcogrammus Pallas. A. (1244)

§ *Boreogadus* Günther.

1563. Pollachius saida Lepechin. G. A. Eu. (1245)

---

[1] *Physiculus dalwigkii* was included in the Synopsis on the basis of an erroneous identification. It should be omitted. A species of *Physiculus* has, however, been recently found. *Physiculus fulvus* Bean, Proc. U. S. Nat. Mus., 1884, 240. Gulf Stream, latitude 40,° in 76 fathoms.

[2] LOTELLA Kaup.

(Kaup, Wiegmann's Archiv, 1858, 88; type *Lotella schlegeli* Kaup.)
This genus differs from *Physiculus* chiefly in the presence in both jaws of an outer row of large teeth. Deep sea. (Name, a diminutive of *Lota*.)
Lotella maxillaris Bean, Proc. U. S. Nat. Mus., 1884, 241. Gulf Stream, latitude 40°.

[3] It seems best to regard the different sections of *Gadus*, as given in the Synopsis, as distinct genera. *Melanogrammus*, especially, is well distinguished by the swollen form of the bones of the shoulder girdle.

[4] For description of *Gadus ogac*, which is regarded by Mr. Dresel as a valid species, see Dresel, Proc. U. S. Nat. Mus., 1884, 246.
(*Gadus ogac* Richardson, Fauna Bor.-Amer., III, 1836, 246. Greenland.)

[5] *Pleurogadus* Bean, nom. gen. nov. to be substituted for *Tilesia*, preoccupied. Type *Gadus navaga* Kölreuter = *Gadus gracilis* Tilesius. (*Bean.*)

535.—HYPSICOMETES Goode. B. (444)

1564. **Hypsicometes gobioides** Goode. B. (1246)

536.—MERLUCIUS Rafinesque. (445)

1565. **Merlucius bilinearis** Mitchill. N. (1247)
1566. **Merlucius merlucius** Linnæus. G. Eu. (1248)
1567. **Merlucius productus** Ayres. C. (1249)

Family CXLVI.—MACRURIDÆ. (121)

537.—MACRURUS Bloch. (447)

1568. **Macrurus berglax**[1] Lacépède. G. Eu. B. (1251)
1569. **Macrurus acrolepis**[2] Bean. A.
1570. **Macrurus carminatus** Goode. B. (1252)
1571. **Macrurus bairdii** Goode & Bean. B. (1253)
1572. **Macrurus asper**[3] Goode & Bean. B.

538.—CORYPHÆNOIDES Gunner (448)

1573. **Coryphænoides rupestris** Gunner. G. B. (1254)
1574. **Coryphænoides carapinus**[4] Goode & Bean. B.

---

[1] *Macrurus berglax* Lacépède = *Macrurus fabricii* Sundevall. To the synonymy add: (*Macrurus berglax* Lacépède, Hist. Nat. Poiss., based on *Macrurus rupestris* Bloch, not of Gunner; the synonymy confused with that of *Coryphænoides rupestris*, which is called "Berglax" ("Rock-Salmon") by Ström.

[2] *Macrurus acrolepis* Bean.

Form of *M. berglax*; width of head ⅗ its height; interorbital width ⅘ eye, which is equal to length of snout, and nearly 4 in head; snout moderate, pointed; maxillary a little more than ⅓ head; second ray of dorsal serrated; distance of anal from snout 2½ in body; pectoral nearly half head; ventral 8 in total length. Head, 4⅘. Depth, 7. D. II, 11, III +. A. 94 +; 7 rows of scales between lateral line and front of dorsal. L. 2½ feet. Straits of Juan de Fuca. A specimen obtained from the stomach of a seal by Mr. J. G. Swan. (*Bean.*)

(Bean, Proc. U. S. Nat. Mus., 1883, 362.)

[3] *Macrurus asper* Goode & Bean.

Dark reddish brown, the spinules with a metallic luster; stouter than in *M. bairdii*; scales small, strong, their free portions covered with vitreous spines in about 7 rows, the middle row not forming a keel, though projecting backward most strongly; interorbital with a little more than length of eye, 4½ in head; snout triangular, depressed; upper ridge prominent anteriorly, ending in advance of concavity of interorbital space; lateral ridges prominent, continued behind the eye; barbel shorter than eye; cleft of mouth reaching to below posterior margin of orbit; second spine of dorsal nearly two-thirds head, not reaching front of soft dorsal when depressed; anal three times as high as second dorsal; vent at a distance from ventral much greater than length of ventral. D. II, 8–105. A. 110. P. 20. V. 10. Scales 7–150–18. Gulf Stream, south of New England.

(Goode & Bean, Bull. Mus. Comp. Zoöl., Vol. X, No. 5, 1883, 196.)

[4] *Coryphænoides carapinus* Goode & Bean.

Scales oval, membranous, without armature, rather large, 22 to 24 in a transverse series. Second ray of dorsal compressed and serrate, as long as head; soft dorsal inserted on a lump-like elevation of the back. Vent nearly below end of first dorsal. Snout acute, projecting beyond the mouth a distance equal to diameter of eye, which is about 4 in head. Bones of head very soft and flexible; surface of head very irreg-

539.—CHALINURA[1] Goode & Bean.
1575. Chalinura simula Goode & Bean. B.

Order AA.—HETEROSOMATA. (U)

Family CXLVII.—PLEURONECTIDÆ. (122)

540.—BOTHUS Rafinesque. (449)
1576. Bothus maculatus Mitchill. N. (1255)
541.—PLATOPHRYS[2] Swainson.
1577. Platophrys leopardinus[3] Günther. P.
1578. Platophrys nebularis[4] Jordan & Gilbert. S.

ular; a very prominent subocular ridge; a prominent ridge from tip of snout to middle of interorbital space; a curved ridge from front of eye above to a point on side of snout just behind its tip. Maxillary extending to opposite posterior margin of pupil, its length half head without snout. Interorbital space equal to length of upper jaw. Head 6. D. 11, 8-100. A. 117. V. 10. Gulf Stream, lat. 40°, in deep water. (*Goode & Bean.*)
(Goode & Bean, Bull. Mus. Comp. Zoöl., Vol. X, No. 5, 197, 1883.)
[1] CHALINURA Goode & Bean.
(Goode & Bean, Bull. Mus. Comp. Zoöl., Vol. X, No. 5, 1883, 198; type, *Chalinura simula.*)
Scales cycloid, fluted longitudinally, with slightly radiating striæ. Snout long, broad, truncate, not much produced. Mouth lateral, subterminal, very large. Head without prominent ridges except the subocular ones and those upon the snout. Suborbital ridge not reaching angle of preopercle. Teeth in the upper jaw in a villiform band, those of the outer series much enlarged, those of the lower jaw uniserial, large. No teeth on vomer or palatines; small pseudobranchiæ present. Gill-rakers spiny, strong, depressible, in double series on anterior arch. Gill membranes apparently free from the isthmus. Ventrals below the pectorals; chin with a barbel. Vertical fins as in *Coryphænoides*. Deep-sea fishes. ($X\dot{\alpha}\lambda\iota\nu\acute{o}\varsigma$, rein; $o\upsilon\rho\acute{a}$, tail.)

Chalinura simula Goode & Bean.

Form of *Coryphænoides*. Snout broad, obtuse, scarcely projecting beyond the mouth; its width at the tip nearly equal to its own length or to the interorbital width. Eye 5 in head, as long as snout; preopercle emarginate behind. Second spine of dorsal serrate; ventral prolonged in a filament which reaches 16th ray of anal. Head 5¼; depth 6⅜. D. II, 9-113. A. 118. P. 20. V. 9. Gulf Stream, about latitude 40°. (*Goode & Bean.*)
(Goode & Bean, l. c., 1883, 199.)

[2] PLATOPHRYS Swainson.

(*Rhomboidichthys* Bleeker).

(Swainson, Nat. Hist. Class'n Fishes, etc., 1839, II, 302; type *Rhombus ocellatus* Agassiz.)
Eyes and color on the left side. Body ovate, strongly compressed; mouth of the large type, but comparatively small; the maxillary one-third or less of the length of the head; teeth small, subequal, in one or two series; no teeth on vomer or palatines. Interorbital space broad and concave, usually broadest in adult males. Gill-rakers moderate. Dorsal fin beginning in front of eye; all its rays simple; ventral of colored side on ridge of abdomen; caudal convex behind; pectoral of left side usually with one or more filamentous rays, longest in the male. Scales very small (in American species); lateral line with a strong arch in front. Coloration usually variegated. Species numerous in warm seas. ($\Pi\lambda\alpha\tau\upsilon\varsigma$, broad; $\omega\varphi\rho\upsilon\varsigma$, eyebrow.)

[3] *Rhomboidichthys leopardinus* Günther. IV, 31; *Parophrys leopardinus* Jordan & Gilbert, Proc. U. S. Nat. Mus., 1884, 260. Guaymas.

[4] *Platophrys nebularis* Jordan & Gilbert, Proc. U. S. Nat. Mus., 1884, 31. Key West, (Jordan); Long Island (Bean).

542.—CITHARICHTHYS Bleeker.

§ *Aramaca*[1] Jordan & Goss.

1579. **Citharichthys ocellatus** Poey. W. (1256 b.)
1580. **Citharichthys pætulus** Goode & Bean. W. (1256)

§ *Hemirhombus* Bleeker.

1581. **Citharichthys ovalis**[2] Günther. P.

§ *Citharichthys.*

1582. **Citharichthys panamensis**[3] Steindachner. P.
1583. **Citharichthys sordidus** Girard. C. (1257)
1584. **Citharichthys stigmæus** Jordan & Gilbert. C. (1257 b.)
1585. **Citharichthys spilopterus** Günther. S. W. P. (1258)
1586. **Citharichthys macrops** Dresel. S.
1587. **Citharichthys arctifrons** Goode. B. (1259)
*1588. **Citharichthys unicornis** Goode. B. (1260)
1589. **Citharichthys microstomus**[4] Gill. N. (1261)

543.—ETROPUS Jordan & Gilbert. (461)

1590. **Etropus crossotus** Jordan & Gilbert. S. P. (1296)

544.—HIPPOGLOSSUS Cuvier. (451)

1591. **Hippoglossus hippoglossus** Linnæus. N. G. A. Eu. (1261)

545.—REINHARDTIUS[5] Gill. (452)

1592. **Reinhardtius hippoglossoides** Walbaum. G. (1262)

546.—ATHERESTHES Jordan & Gilbert. (453)

1593. **Atheresthes stomias** Jordan & Gilbert. C. A. (1263)

547.—PARALICHTHYS Girard. (454)

1594. **Paralichthys adspersus**[6] Steindachner. P.
1595. **Paralichthys californicus** Ayres. C. (1264)

---

[1] *Aramaca* Jordan & Goss, sub-genus nova, type *Hemirhombus pætulus* Bean. This group includes species which have the broad, concave interorbital space, elongate pectorals, and other characters of *Platophrys*, but are without arch in the lateral line, as in *Hemirhombus* and *Citharichthys.*

[2] *Hemirhombus oralis* Günther, Proc. Zoöl. Soc. London, 1864, 154; Günther, Fishes Centr. Amer., 1869, 472. Mazatlan to Panama.

[3] *Citharichthys panamensis* Steindachner, Ichth. Beiträge, III, 62, 1875. Mazatlan to Panama.

[4] *Citharichthys microstomus* Gill, Proc. Ac. Nat. Sci. Phila., 1864, 223. Atlantic coast. This species, lately rediscovered by Dr. Bean, is distinct from *C. spilopterus*, having a considerably smaller mouth. It approaches *E. crossotus*, but the latter species has the mouth still smaller and the body deeper.

[5] *Reinhardtius* Gill, has priority over *Platysomatichthys*, but was proposed without definition or explanation.

[6] *Paralichthys adspersus* Steindachner, Ichth. Notizen, V. 1867-9. Mazatlan to Peru.

REPORT OF COMMISSIONER OF FISH AND FISHERIES. [134]

1596. Paralichthys dentatus [1] Linnæus. N. S. (1265)
1597. Paralichthys lethostigma [2] Jordan & Gilbert. N. S. (1266)
1598. Paralichthys albigutta Jordan & Gilbert. S. (1267)
1599. Paralichthys squamilentus Jordan & Gilbert. S. (1268)
1600. Paralichthys oblongus Mitchill. N. (1269)

548.—ANCYLOPSETTA [3] Gill.

1601. Ancylopsetta quadrocellata Gill. S. (1270)
1602. Ancylopsetta dilecta [4] Goode & Bean. B.

---

[1] *Paralichthys dentatus* (L.) *Common Spotted Flounder, Northern Flounder.*
  Cape Cod to Florida, most abundant northward. The description in the synopsis (p. 822) of *P. ophryas*, belongs here. From *P. lethostigma*, it is especially distinguished by the more numerous (5 + 14) gill-rakers, and by the much more spotted coloration. The interorbital space is also narrower in specimens of the same size.
  (*Pleuronectes dentatus* L., Syst., Nat., Ed. XII, 1766, 458, from a specimen from Dr. Garden; this specimen has been examined by Dr. Bean; it belongs to the present species; *Pleuronectes melanogaster* Mitchill, Trans. Lit. & Phil. Soc. N. Y., 1815, 1, 390; *Platessa ocellaris* DeKay, New York Fauna, Fishes. 1842, 300; *Paralichthys ophryas* Jor. & Gilb., Syn. Fish. N. A., 822; *Paralichthys ocellaris* Jor. & Gilb., l. c., 972, and Proc. U. S. Nat. Mus. 1882, 617; *Pseudorhombus ocellaris* Günther, IV, 430.)

[2] *Paralichthys lethostigma* Jordan and Gilbert.
  Cape Cod to Florida and Texas, most abundant southward. Darker and more uniform in color than the true *dentatus*, the gill-rakers smaller and fewer (2 + 10) and the interorbital space broader.
  (*Platessa oblonga* DeKay, New York, Fauna, Fish., 1842, 299, not *Pleuronectes oblongus* Mitchill; *Pseudorhombus dentatus* and *oblongus* Günther, IV, 425, 426; *Paralichthys dentatus* Jor. & Gilb., Synopsis 822, and Proc. U. S. Nat. Mus. 1882, 617; *Paralichthys lethostigma* Jordan & Gilbert, Proc. U. S. Nat. Mus. 1884, 237. The original type of *P. dentatus* examined by Dr. Bean in London proves to belong to the species having numerous gill-rakers.

[3] It seems more natural to regard *Ancylopsetta* and *Xystreurys* as genera distinct from *Paralichthys*. *Notosema* Goode & Bean (*dilecta*) seems scarcely different from *Ancylopsetta*.

[4] *Ancylopsetta dilecta* (Goode & Bean).
  Dark brown, speckled with darker; three large, subcircular ocellated spots, nearly as large as eye, with white center, dark iris, narrow dark margin, and a brown encircling outline. These spots arranged in an isosceles triangle, the apex on the lateral line, the others distant from the lateral line a distance equal to their own diameter; the lower near tip of ventral. Fins blotched with darker brown. Right side white. Body elliptical, the caudal fin peduncunlate; mouth moderate, the maxillary 2½ in head; teeth uniserial, those in front much largest. Eye large, 3 in head, the interorbital space very narrow. Gill-rakers subtriangular, moderately numerous. Pectoral fins unequal, the left 5½ in body. Ventral of colored side much produced, more than three times length of right ventral. First eight rays of dorsal exserted, forming a somewhat separate division, the second and third longest half greatest depth of body. Scales small, highly ctenoid. Head 3½; depth 2. D. 69; A. 76; P. 11; V. 6; lat. l. 48 (in straight portion). Gulf Stream, off the Carolina coast. (Goode & Bean.)
  (*Notosema dilecta* Goode & Bean, Bull. Mus. Comp. Zoöl., XIX, 193.)
  The genus *Notosema* is distinguished from *Paralichthys* "on account of its elongated ventral fin, the triangular elongation of the anterior rays of the dorsal and the highly ctenoid character of the scales on the colored side of the body." These characters are all, however, of degree only, and all exist in *Ancylopsetta quadrocellata*.

[135] CATALOGUE OF THE FISHES OF NORTH AMERICA.

549.—XYSTREURYS Jordan & Gilbert.

1603. Xystreurys liolepis Jordan & Gilbert. C. (1271)

550.—HIPPOGLOSSINA[1] Steindachner. (455)

1604. Hippoglossina macrops Steindachner. P.

551.—HIPPOGLOSSOIDES Gottsche. (456)

§ *Eopsetta*[2] Jordan & Goss.

1605. Hippoglossoides jordani Lockington. C. (1274)

§ *Hippoglossoides*.

1606. Hippoglossoides platessoides Fabricius. N. G. Eu. (1272)
1607. Hippoglossoides elassodon Jordan & Gilbert. C. A. (1273)

§ *Lyopsetta*[3] Jordan & Goss.

1608. Hippoglossoides exilis Jordan & Gilbert. C. A. (1275)

552.—PSETTICHTHYS Girard.

1609. Psettichthys melanostictus Girard. C. (1276)

553.—PLEURONICHTHYS Girard. (456)

1610. Pleuronichthys decurrens Jordan & Gilbert. C. (1277)
1611. Pleuronichthys verticalis Jordan & Gilbert. C. (1278)
1612. Pleuronichthys cœnosus Girard. C. A. (1279)

554.—HYPSOPSETTA Gill. (457)

1613. Hypsopsetta guttulata Girard. C. (1280)

555.—PAROPHRYS Girard.

1614. Parophrys vetulus Girard. C. A. (1281)

556.—ISOPSETTA Lockington.

§ *Isopsetta*.

1615. Isopsetta isolepis Lockington. C. (1282)

---

[1] HIPPOGLOSSINA Steindachner.
(Steindachner, Ichth. Beitr. V, 13, 1876; type *Hippoglossina macrops* Steindachner.) This genus is very close to *Paralichthys*, differing chiefly in the dentition, the teeth eing small and uniform in size, arranged in a single row. The scales are ctenoid. 'he eyes are unusually large in the single known species, which bears a remarkable resemblance to *Hippoglossoides jordani*. The lateral line is however anteriorly arched in *Hippoglossina*, but straight in the latter species. (Name a diminutive of *Hippoglossus*.)
Hippoglossina macrops Steindachner, l. c. Mazatlan, probably from rather deep water.

[2] *Eopsetta* Jordan & Goss, subgenus nova, for *Hippoglossoides jordani* Lockington ἠώς, excellent; ψῆττα, flounder), characterized by the biserial upper teeth and by other peculiarities.

[3] *Lyopsetta* Jordan & Goss, subgenus nova, for *Hippoglossoides exilis* Jordan & Gilbert (λύω, to loosen; ψῆττα, flounder), characterized by the large, loose scales, biserial upper teeth, and feeble structure.

§ *Inopsetta*[1] Jordan & Goss.

1616. Isopsetta ischyra Jordan & Gilbert. A. (1283)

557.—LEPIDOPSETTA Gill.

1617. Lepidopsetta bilineata Ayres. C. A. (1284)

558.—LIMANDA Gottsche.

1618. Limanda ferruginea Storer. N. (1285)
1619. Limanda aspera Pallas. A. (1286)
1620. Limanda beani Goode. B. (1287)

559.—PLEURONECTES[2] Linnæus. (458)

§ *Platichthys* Girard.

1621. Pleuronectes stellatus Pallas. A. C. (1288)

§ *Pleuronectes*.

1622. Pleuronectes quadrituberculatus Pallas. A. (1289)
1623. Pleuronectes glaber Storer. N. (1290)
1624. Pleuronectes glacialis Pallas. A. (1291)

§ *Pseudopleuronectes* Bleeker.

1625. Pleuronectes americanus Walbaum. N. (1292)

560.—GLYPTOCEPHALUS Gottsche. (459)

1626. Glyptocephalus cynoglossus Linnæus. N. Eu. B. (1293)
1627. Glyptocephalus zachirus Lockington. C. (1294)

561.—CYNICOGLOSSUS Bonaparte. (460)

1628. Cynicoglossus pacificus Lockington. C. A. (1295)

562.—DELOTHYRIS[3] Goode. (462)

1629. Delothyris pellucidus Goode. B. (1296)

563.—MONOLENE Goode. (463)

1630. Monolene sessilicauda Goode. B. (1298)

---

[1] *Inopsetta* Jordan & Goss, subgenus nova, type *Parophrys ischyrus* Jordan & Gilbert. (Ἴς, sinew; ψῆττα, flounder.) This fish is allied to *Pleuronectes stellatus*, but has an accessory dorsal branch to the lateral line as in *Isopsetta isolepis*, from which it differs in form, and in the rough, loosely imbricated scales.

[2] The genus *Pleuronectes* as retained in the Synopsis, is unnatural, species very diverse in their characters being retained in it. I have, therefore, here recognized its chief constituents as distinct genera. *Parophrys*, *Isopsetta*, *Lepidopsetta*, and *Limanda* seem certainly worthy of such recognition. Possibly *Platichthys*, *Inopsetta* and *Pseudopleuronectes*, also, are worthy of such retention.

[3] DELOTHYRIS Goode.

(Goode, Proc. U. S. Nat. Mus. 1883, 110; type *Thyris pellucidus* Goode; name a substitute for *Thyris*, preoccupied; δῆλος, clear; θυρίς, window.) We have no doubt that this is a larval form possibly of some fish as yet unknown, allied to *Citharichthys*. Small transparent flounders having all the characters of *Delothyris*, but less elongate than *D. pellucidus*, have been taken by the writer at Key West. These are thought to be larvæ of some *Platophrys* or *Citharichthys*.

Family CXLVIII.—SOLEIDÆ. (123)

564.—ACHIRUS Lacépède. (464)

§ *Bæostoma* [1] Bean.

1631. Achirus brachialis Bean. S. (1299 c.)
1632. Achirus comifer [2] Jordan & Gilbert. W.
1633. Achirus mazatlanus [3] Steindachner. P.
1634. Achirus inscriptus [4] Gosse. W.

§ *Achirus*.

1635. Achirus achirus [5] Linnæus. W. S. (1299 b.)
1635 b. *Achirus achirus mollis* Mitchill. N. (1299)

565.—APHORISTIA Kaup. (465)

1636. Aphoristia atricauda Jordan & Gilbert. C. (1300)
1637. Aphoristia plagiusa Linnæus. S. (1301)
1638. Aphoristia nebulosa [6] Goode & Bean. B.

---

[1] *Bæostoma* should probably be regarded as a subgenus of *Achirus* rather than as a distinct genus. Among the numerous species, the pectoral of the right side is found in every degree of development. In some species, a small pectoral is found on the left side in some specimens, while it is wanting in others. Still other species have also two pectorals developed.

[2] *Achirus comifer* Jordan & Gilbert, Proc. U. S. Nat. Mus., 1884, 31. Key West.

[3] *Solea mazatlana* Steindachner, Ichth. Notizen, IX, 1869, 23 (July) = *Solea (Monochir) pilosa* Peters, Berliner Monatsber., 1869, 709 (August). Mazatlan, southward.

[4] *Achirus inscriptus* Gosse.

Olivaceous, covered with an irregular network of blackish lines; this network rather finer on the head; some specimens crossed by irregular but nearly straight vertical lines; others without traces of these; dorsal and anal colored like the body, rather darker, with a paler edge; caudal abruptly whitish, immaculate; blind side immaculate, darker on the fins; hair-like appendages whitish; scales about head enlarged and fringed, especially on blind side; lip of eyed side much fringed: interorbital width less than eye; upper eye slightly in advance of lower; right pectoral of three rays, the middle one somewhat longer than the others; left ventral of one or two very small rays often entirely absent; right side with scattered cilia, which are mostly whitish; ventrals 5-rayed, the right ventral joined to the anal; head, 3¾; depth, 1¾; D., 54; A., 40; lat. l., 75 to 80. West Indies, north to Key West.

(*Achirus inscriptus* Gosse, Naturalist's Sojourn Jamaica, 52; *Solea inscripta* Günther, IV, 473; *Monochir reticulatus* Poey, Memorias Cuba, II, 1861, 317; *Solea reticulata* Günther, IV, 472; *Achirus inscriptus* Jordan, Proc. U. S. Nat. Mus., 1884, 143.)

[5] The name *Pleuronectes achirus* L. (*Achirus fasciatus* Lac.) was based on specimens from Surinam; the name *Pleuronectes lineatus* on the figures of Brown and Sloane of fishes from Jamaica. If, therefore, the West Indian form is considered distinct from the northern one, the former must be *Achirus achirus* or *Achirus lineatus*, and the latter must take Mitchill's name, "*mollis*." If considered as varieties of one species, the West Indian form has the prior names.

[6] *Aphoristia nebulosa* Goode & Bean.

Grayish, everywhere mottled with brown; median keel on each scale dark and prominent. Body comparatively slender; scales small, rough; jaws and snout naked; interorbital space with one row of scales. Teeth small, apparently equally developed on both sides. Ventral well separated from anal, its longest ray 3 in head. Head 5¾; depth 4¾, D. 119, A.107, P. O. V. 5. Scales 120-50. L. 3¼ inches. Gulf Stream, off the coast of Carolina. (*Goode & Bean*.)

(Goode & Bean, Bull. Mus. Comp. Zoöl., XIX, 1883, 192).

Order BB.—PEDICULATI. (V.)

Family CXLIX.—LOPHIIDÆ. (124)

566.—LOPHIUS Linnæus. (466)

1639. Lophius piscatorius Linnæus. N. Eu. (1302)

Family CL.—ANTENNARIIDÆ. (125a.)

567.—PTEROPHRYNOIDES Gill. (466 b.)

1640. Pterophrynoides histrio Linnæus. S. O. (1303)

568.—ANTENNARIUS Lacépède. (467)

1641. Antennarius annulatus Gill. W. (1304)
1642. Antennarius ocellatus[1] Bloch & Schneider. W. (1305)
1643. Antennarius sanguineus[2] Gill. P.
1644. Antennarius strigatus Gill.[3] P.

569.—CHAUNAX Lowe. (468)

1645. Chaunax pictus Lowe. B. (1306)

Family CLI.—CERATIIDÆ. (125 b.)

570.—CERATIAS Kröyer. (469)

1646. Ceratias holbölli Kröyer. B. G. (1307)

'571.—MANCALIAS[4] Gill. (470)

1647. Mancalias uranoscopus Murray. B. (1308)

---

[1] *Lophius vespertilio* Var. *ocellatus* Bloch & Schneider, Syst. Ichth., 1801, 142, based on the Pescador of Parra = *Antennarius ocellatus* Poey, Syn. Pisc. Cub., 1868, 105 = *Antennarius pleurophthalmus* Gill.

[2] *Antennarius sanguineus* Gill, Proc. Ac. Nat. Sci. Phila., 1863, 91 = *Antennarius leopardinus* Günther, Proc. Zoöl. Soc., London, 1864, 151. Cape San Lucas to Panama.

[3] *Antennarius strigatus* Gill, l. c. 92 = *Antennarius tenuifilis* Günther, Fish Centr Amer. 1869, 440 = *Antennarius strigatus* Jordan & Gilbert, Proc. U. S. Nat. Mus., 1882, 630. Cape San Lucas to Panama.

[4] The following notes on fishes similar to *Mancalias* were published in Forest and Stream of Nov. 8, 1883, by Dr. Theodore Gill:

"*Typhlopsaras.*—Ceratiines with an elongated trunk, rectilinear back, obsolete or no eyes, far exserted basal joint of the anterior spine and shortened terminal joint, a small intermediate and a pair of pedunculated dorsal appendages some distance in advance of the dorsal fin, and reduced pectoral fin with about 5 or 6 rays.

"*Typhlopsaras shufeldti.*—The first joint of the rod-like spine reaches to the axil of the dorsal fin, and the bulb to the base of the caudal fin, when the spine is bent backward; the bulb is pear-shaped and without any appendages; the dorsal has 4 rays, the anal 4, the caudal 8 (the median, 4 of which are forked), and there are 4 or 5 pectoral rays. A single specimen was found. I have dedicated the species to my esteemed friend, Dr. R. W. Shufeldt, U. S. A., the well-known ornithotomist.

"The name *Typhlopsaras* is a compound from the Greek *tuphlos* (blind) and *psaras* (angler), meaning 'blind angler.'

"*Cryptopsaras.*—Ceratiines with shortened trunk, longitudinally convex back, small but conspicuous eyes, concealed basal joint of the anterior spine and elongated ter-

572.—ONEIRODES Lütken. (471)

1648. Oneirodes eschrichti Lütken. B. G. (1309)

573.—HIMANTOLOPHUS Reinhardt. (472)

1649. Himantolophus grœnlandicus Reinhardt. B. G. (1310)
1650. Himantolophus reinhardti Lütken. B. G. (1311)

Family CLII.—MALTHIDÆ. (126)

574.—MALTHE Cuvier. (473)

1651. Malthe vespertilio Linnæus. S. W. (1312)
1651b. *Malthe vespertilio radiata*[1] Mitchill. S. (1313)
1652. Malthe elater[2] Jordan & Gilbert. P.

575.—HALIEUTICHTHYS Poey. (474)

1653. Halieutichthys aculeatus Mitchill. W. (1314)

576.—HALIEUTÆA Cuvier & Valenciennes. (475)

1654. Halieutæa senticosa Goode. B. (1315)

Order CC.—PLECTOGNATHI. (W.)

Family CLIII.—OSTRACIIDÆ. (476)

577.—OSTRACION Linnæus. (476)

§ *Lactophrys.* Swainson.

1655. Ostracion triquetrum Linnæus. W. (1316 b.)
1656. Ostracion trigonum Linnæus. W. (1316)
1657. Ostracion tricorne[3] Linnæus. W. S. (1317)

---

minal joint, a large intermediate globular and a pair of sub-pedunculated lateral dorsal appendages near the front of the dorsal fin, and well-developed pectorals of about 15 rays.

"*Cryptopsaras couesii.*—The basal joint of the rod-like spine is almost entirely concealed and procumbent, and the distal joint alone free, reaching backward to the dorsal tubercles; the bulb is pyriform and surmounted by a long whitish filament; the dorsal and anal have each 4 spines, the caudal 8 (the 4 middle dichotomous), and the pectorals each about 15 rays. The species has been named after the eminent ornithologist, Dr. Elliott Coues. The name is derived from the Greek *cruptos* (concealed,) and *psaras* (fisherman), and has reference to the concealed 'rod' or basal joint of the anterior spine or fishing apparatus."

[1] *Malthe cubifrons* Rich., seems to be only an extreme variety of *Malthe vespertilio*. Every gradation in size and form of the rostral process exists between the very long-nosed var. *longirostris*, to the button-nosed *cubifrons*, and thus far I am unable to show any dividing lines. The original record of *Malthe cubifrons* as from Labrador was an error. It is not certainly known from any point north of Florida. The name *Lophias radiatus* Mitchill, Amer. Monthly Mag., March, 1818, 326, is prior to that of *cubifrons*. The short-snouted form may therefore stand as—

Malthe vespertilio radiata. (See Jordan & Swain, Proc. U. S. Nat. Mus., 1884, 234.)

[2] *Malthe elater* Jordan & Gilbert, Proc. U. S. Nat. Mus., 1881, 365. Mazatlan.

[3] *Ostracion tricornis* Linnæus. Syst, Nat, X, 1758, 331 = *Ostracion quadricornis* Linnæus, (lower down on the same page.)

Family CLIV.—BALISTIDÆ.

578.—BALISTES Linnæus. (477)

1658. Balistes vetula Linnæus. W. (1318)
1659. Balistes carolinensis[1] Gmelin. S. W. Eu. (1319)
1660. Balistes powelli Cope. Acc. (1320)
1661. Balistes polylepis[2] Steindachner. P.
1662. Balistes capistratus[3] Shaw. P.

579.—MONACANTHUS Cuvier. (478)

§ *Monacanthus.*

1663. Monacanthus ciliatus[4] Mitchill. W. (1321, 1323)
1664. Monacanthus hispidus Linnæus. S. N. (1322)
1665. Monacanthus spilonotus Cope. W. (1324)

§ *Cantherhines* Swainson.

1666. Monacanthus pullus Ranzani. W. (1325)

580.—ALUTERA Cuvier. (479)

1667. Alutera schœpfi Walbaum. N. S. (1326)
1668. Alutera scripta Osbeck. W. (1327)

Family CLV.—TETRODONTIDÆ.

581.—LAGOCEPHALUS Swainson. (480)

1669. Lagocephalus lævigatus Linnæus. W. S. (1328)

582.—TETRODON[5] Linnæus. (481)

1670. Tetrodon politus Girard. C. P. (1329)
1671. Tetrodon testudineus Linnæus. W. (1330.)

---

[1]*Balistes carolinensis* Gmelin, Syst. Nat., 1788, 1468 (as variety of *B. vetula*). *Balistes capriscus* Gmelin occurs first on page 1471, and is based on a confusion of several species. *Balistes powelli* is possibly the young of this species.

[2]*Balistes polylepis* Steindachner, Ichth. Beitr., V, 21, 1876. Mazatlan to Panama.

[3]*Balistes capistratus* Shaw, Gen. Zoöl., V, 417, 1804 (based on *Baliste bridé* Lacépède)= *Balistes mitis* Bennett = *Balistes frenatus* Richardson. Mazatlan to Panama.

[4]*Balistes ciliatus* Mitchill, Amer. Monthly Mag., 1818, 326 = *Monacanthus occidentalis* Günther = *Monacanthus davidsoni* Cope. See Jordan, Proc. U. S. Nat. Mus., 1884, 145.

[5]The earliest attempt at subdivision of the genus *Tetrodon* as left by Cuvier seems to be that of Swainson. In his restricted genus *Tetrodon* no Linnæan species are retained, his "*Tetrodon testudineus*" being that of Bloch, not of Linnæus. The next attempt is that of Müller, who did not retain the name *Tetrodon* for any of his subdivisions. The next attempt at subdivision seems to be that of Bleeker, who retained the name *Tetrodon*, in accordance with his custom, for the first species mentioned by Linnæus, *T. testudineus*. This seems to me the earliest use of the restricted name *Tetrodon* which can stand.

In a recent paper, Dr. Gill (Proc. U. S. Nat. Mus., 1884, 420) has adopted a different view. The *Tetrodon* of Swainson contains three species congeneric with one of the Linnæan species (*lineatus*). This species belongs to Müller's genus *Arothron*, and to *Arothron* Dr. Gill transfers the name *Tetrodon*, reserving for the *Tetrodon* of Bleeker and of our Synopsis the name *Cirrhisomus* of Swainson.

1671b. *Tetrodon testudineus annulatus*[1] Jenyne. P.
1672. **Tetrodon spengleri** Bloch. W. (1331)
1673. **Tetrodon nephelus**[2] Goode & Bean. S.W. (1332 b.)
1674. **Tetrodon turgidus** Mitchill. N. (1332)
1675. **Tetrodon trichocephalus** Cope. Acc. (1333).

583.—PSILONOTUS[3] Swainson.

1676. **Psilonotus punctatissimus** Günther. P.

Family CLVI.—DIODONTIDÆ.

584.—TRICHODIODON Bleeker. (482)

1677. **Trichodiodon pilosus** Mitchill. O. (1334)

585.—DIODON Linnæus. (483)

1678. **Diodon hystrix** Linnæus. W.P. (1335)
1679. **Diodon liturosus** Shaw. W.P. (1336)

586.—CHILOMYCTERUS (Bibron) Kaup. (484)

1680. **Chilomycterus geometricus** Mitchill. N.S. (1337)
1681. **Chilomycterus fuliginosus** DeKay. N. (1337 b.)
1682. **Chilomycterus reticulatus** Linnæus. W. (1337 c.)

Family CLVII.—ORTHAGORISCIDÆ. (130)

587.—MOLA[4] Cuvier. (485, 486)

1683. **Mola mola** Linnæus. N.S.W. O.C.Eu.P. (1338, 1339)

---

[1] *Tetrodon annulatus* Jenyns, Zoöl. Beagle, 1842, 153 = *Tetrodon heraldi* Günther, VIII, 283. Gulf of California to Peru. This species is little, if at all, different from *T. testudineus*.

[2] *Tetrodon nephelus* is extremely variable in regard to its spinous armature. Specimens from Key West show all gradations from entire smoothness above and below to the condition described in the text (page 966). Older specimens are generally less prickly than young ones.

[3] PSILONOTUS Swainson.
(*Anosmius* Peters; *Tropidichthys* and *Canthogaster* Bleeker; *Anchisomus* Richardson.) (Swainson, Nat. Hist. Classn. Anim., II, 1839, 328; type *Tetrodon rostratus* Bloch.)
This genus differs externally from *Tetrodon* in having the nostrils obsolete, and the back compressed to a keel. The skeleton differs so widely from that of *Tetrodon* that Dr. Gill (Proc. U.S. Nat. Mus., 1884, 422) has proposed to regard it as forming a distinct family, *Psilonotidæ*. Species rather numerous in the tropics. (Ψιλος, bare; νῶτος, back.)
Psilonotus punctatissimus Günther. *Tetrodon punctatissimus* Günther, VIII, 302 = *Tetrodon oxyrhynchus* Lockington, Proc. Ac. Nat. Sci. Phila., 1881, 116. Gulf of California to Panama.

[4] The generic name *Mola* first appears in Cuvier, Tableau Élémentaire, 1798, p. 423, thus having three years priority over *Orthagoriscus* (1801).
The recent researches of Mr. John A. Ryder render it very probable that the small fishes known as *Molacanthus* are, after all, young forms of *Mola*. I therefore omit *Molacanthys nummularis*.
Ranzania truncata (No. 1139 b) should not be included in the present list, as it has not been taken nearer our coast than the Bermuda Islands.

RECAPITULATION.

The following is an approximate statement of the number of species and subspecies, now known, belonging to each of the principal faunal areas. No species is counted twice, but in case of the numerous species which range over several faunal areas each is referred to that area which is supposed to be most properly its home, or to that in which its occurrence has been longest known. In regard to many species such an assignment is simply arbitrary, and in this fact lies the chief element of error in the following list. Thus many Arctic shore fishes belong to the Bassalian fauna of New England, while many West Indian species occur northward more or less frequently as far as Cape Cod. No faunal region on our coast is bounded by sharp lines:

	Species.
Bassalian or deep-sea fauna of the Atlantic	105
Arctic (Greenland) fauna	65
New England (Newfoundland to Cape Hatteras)	95
South Atlantic and Gulf coast (shore fauna)	140
West Indian fauna (including Florida Keys and "Snapper Banks" of Pensacola)	290
Tropical fauna of the Pacific (Gulf of California, southward)	240
Californian fauna (Cape Flattery to Cerros Island)	220
Alaska (Cape Flattery to Bering's Straits)	90
Pelagic species	35
Fresh waters: East of Rocky Mountains	465
Fresh waters: Between Rocky Mountains and Sierra Nevada (Great Basin, &c.)	75
Fresh waters: West of Sierra Nevada and Cascade Range	50
Total	1,870

INDIANA UNIVERSITY,
January 1, 1885.

INDEX.

[NOTE.—Figures in parenthesis refer to the consecutive numbers assigned the genera in their natural order; the page references are to figures in brackets on the inside of the page.]

	Page.
abbreviata, Chimæra	12
Abeona aûrora (1134)	96
minima (1133)	96
abildgaardii, Sparus	101
Acantharchus pomotis (847)	76
acanthias, Squalus (19)	5
Acanthocybium petus	68
solandri (770)	68
Acanthopteri	58
Acanthuridæ (Family CXVII)	103
Acanthurus chirurgus	103
phlebotomus	103
Achirus achirus (1635)	137
mollis (1635 b)	137
brachialis (1631)	137
comifer (1632)	137
fasciatus	137
inscriptus (1634)	137
lineatus	137
mazatlanus (1633)	137
achirus, Achirus (1635)	137
Pleuronectes	137
Acipenser brevirostris (105)	13
medirostris (103)	13
rubicundus (104)	13
sturio oxyrhynchus (101)	13
transmontanus (102)	13
Acipenseridæ (Family XXVII)	13
acipenserinus, Podothecus (1381)	114
ackleyi, Raia (67)	11
Acrochilus alutaceus (199)	20
acrolepis, Macrurus (1569)	131
Actinochir	115
Actinopteri	13
aculeatus var. (1063 b)	91
Gasterosteus (713)	63
Halieutichthys (1653)	139
Stenotomus	91
acuminata, Sciæna	94
acuminatus, Clinus	120
Eques (1093)	94
Ophisurus (617)	53
acuta, Dussumieria	35
acutirostris, Ichthyapus	52
acutum, Hæmulon (1051)	90
Adinia multifasciata (556)	48
adinia, Fundulus (565)	49
adspersus, Ctenolabrus (1150)	97
Paralichthys (1594)	133
adustus, Gobiesox (1415)	116
æglefinus, Melanogrammus (1555)	130

	Page.
Ælurichthys marinus (141)	16
nuchalis	16
panamensis (142)	16
pinnimaculatus (143)	16
æneus, Cottus (1334)	111
ænigmaticus, Icosteus (825)	73
æpypterus, Ammocœtes (7)	4
æquidens, Culius	105
Eleotris (1222)	105
æsculapius, Plagyodus (473)	38
æsopus var. (685 g)	78
æstivalis, Clupea (445)	36
Gobio	29
Hybopsis (340)	29
æthalorus, Carcharhinus (34)	7
Aëtobatis laticeps	12
afer, Epinephelus	84
Gymnothorax	52
affine, Siphostoma (690)	61, 62
affinis, Atherinops (737)	65
Chimæra (98)	12
Cremnobates (1467)	121
Exocœtus	61
Gambusia (588)	50
Gila (361)	30
afra, Muræna	52
agassizii, Alepocephalus (427)	34
Bathysaurus (483)	39, 40
Chologaster (542)	47
Holconotus (1140)	96
aggregatus, Micrometrus (1137)	96
Agnus anoplus	118
Agonidæ (Family CXXV)	113
Agonostomus nasutus (722)	64
telfairi	64
Agonus	113
Agosia carringtoni (325)	28
chrysogaster (322)	28
metallica (323)	28
novemradiata (324)	28
nubila (326)	28
oscula (327)	28
alabamæ, Notropis	27
alalonga, Orcynus (773)	69
alascanus, Ammodytes (748)	66
alatus, Prionotus (1386)	114
albescens, Echeneis	66
Remora (754)	66
albidus, Amiurus (129)	15
Ptychostomus	19
Tetrapturus (758)	67

[143]

	Page.
albigutta, Paralichthys (1598)	134
Albula vulpes (429)	34
albula, Mugil	64
Albulidæ (Family XXXV)	34
albulus, Lepomis (872)	77
album, Moxostoma (182)	19
Alburnellus jemezanus	27
megalops	26
percobromus	27
umbratilis	26
Alburnops	23
blennius	24, 26
illecebrosus	23
saludanus	24
shumardi	23
taurocephalus	22
Alburnus rubellus	27
zonatus	26
alburnus, Menticirrus (1109)	94
alectrolophus, Anoplarchus	122
Alepidosauridæ (Family XLII)	38
Alepidosaurus	38
Alepocephalidæ (Family XXXIV)	34
Alepocephalus agassizii (427)	34
bairdii (426)	34
productus (428)	34
Algansea antica (411)	32
bicolor (408)	32
dimidiata (413)	32
formosa	32
obesa (406)	32
olivacea (412)	32
parovana (409)	32
symmetrica (407)	32
thalassina (410)	32
vittata (414)	32
aliciæ, Phoxinus (399)	31
aliciola, Seriola	72
aliciolus, Trachurus	72
alliteratus, Euthynnus (775)	69
Allosomus	43
Alopias vulpes (48)	9
Alopiidæ (Family XII)	9
Alosa	36
alosoides, Hyodon (430)	34
Alphestes multiguttatus (991)	84
alticus, Salarias	120
altipinnis, Notropis (291)	26
altivelis, Cremnobates (1464)	121
Trachypterus (1212)	104
altus, Chorinemus	72
Oligoplites (812)	72
Pseudopriacanthus (1001)	86
alutaceus, Acrochilus (199)	20
Alutera schœpfi (1667)	140
scripta (1668)	140
alutus, Apogon (1076)	92
Alvarius fonticola (946)	81
lateralis (943)	81
prœliaris (944)	81
punctulatus (945)	81
alvordi var. (1320 e)	111
Alvordius crassus	70
maculatus	79
variatus	79

	Page.
amabilis, Notropis (292)	26
amara, Dionda (209)	21
amarus var. (240 b)	24
Notropis	24, 26
Ambloplites rupestris (845)	76
amblops, Ceratichthys	28
Hybopsis (331)	29
Amblyopsidæ (Family LIII)	47
amblyopsis, Eleotris (1221)	105
Amblyopsis, spelæus (539)	47
amblyrhynchus, Caranx (782)	70
americanus, Ammodytes (747)	66
Amphiprion	83
Apogon	92
Cyprinus	33
Esox (597)	50
Hemitripterus (1300)	100
Histiophorus	67
Istiophorus (759)	67
Pleuronectes (1625)	136
Polyprion (974)	83
Roccus (957)	82
Amia calva (110)	13
retrosella	92
Amiidæ (Family XXIX)	13
Amitra liparina	115
Amitrus	115
Amiurus albidus (129)	15
brachyacanthus	14
brunneus (122)	14
cragini	14
erebennus (128)	15
lophius	15
lupus (130)	15
marmoratus	15
melas (124)	14
natalis (127)	15
bolli (127 c)	15
lividus (127 b)	15
nebulosus (125)	14
catulus (125 b)	14, 15
marmoratus (125 c)	15
nigricans (132)	15
niveiventris (131)	15
obesus	14
platycephalus (123)	14
ponderosus (133)	15
prosthistius	15
vulgaris (126)	15
xanthocephalus	14
Ammocœtes æpypterus (7)	4
aureus (6)	4
borealis	4
cibarius (5)	4
concolor	4
tridentatus (4)	3
Ammocrypta	78
beani (878)	77
clara (879)	77
pellucida (880)	77
vivax (881)	77
Ammodytes alascanus (748)	66
americanus (747)	66
personatus (747 b)	66
dubius (749)	66

CATALOGUE OF THE FISHES OF NORTH AMERICA.

	Page.
Ammodytidæ (Family LXXIX)	66
Amphiprion americanus	83
matejuelo	75
Amphistichus argenteus (1142)	96
ampullaceus, Saccopharynx (648)	57
analigutta, Pomacentrus	102
analis, Holconotus (1138)	96
Lutjanus (1014)	87
Notacanthus (653)	58
Oligocottus (1362)	113
Umbrina	94
analogus, Epinephelus (990)	84
Kyphosus (1070)	92
Pimelepterus	92
analostanus, Notropis	25
Anarrhichadidæ (Family CXXXVIII)	123
Anarrhichas latifrons (1500)	123
lepturus (1501)	123
lupus (1498)	123
minor (1499)	123
Anarrhichthys ocellatus (1502)	123
Anchisomus	141
ancipitirostris, Histiophorus	67
Ancylopsetta dilecta (1602)	134
quadrocellata (1601)	134
Anguilla	52
anguilla rostrata (638)	55
cubana	55
rostrata	55
texana	55
tyrannus	55
anguilla, Anguilla (638)	55
anguillaris, Lumpenus (1490)	123
Zoarces (1503)	124
Anguillidæ (Family LX)	52, 55
anguilliformis, Pholidichthys (1495)	123
Anisotremus	88
bilineatus (1037)	89
cæsius (1035)	89
davidsoni (1038)	89
dovii (1034)	89
interruptus (1036)	89
modestus	89
tæniatus	89
virginicus (1039)	89
tæniatus (1039 b)	89
anisurum, Moxostoma (190)	20
annularis, Pomoxys (842)	76
annulatus var. (1671 b)	141
Antennarius (1641)	138
Tetrodon	141
anogenus, Notropis (227)	23
anolis, Saurus	39
Synodus (481)	39
anomalum, Campostoma (196)	20
Anoplarchus electrolophus	122
atropurpureus (1479)	122
Anoplogaster	74
Anoplopoma fimbria (1261)	107
anoplos, Uranoscopus	118
auoplus, Agnus	118
Astroscopus (1430)	117, 118
Anosmius	141
Antennariidæ (Family CL)	138
Antennarius annulatus (1641)	138

	Page.
Antennarius leopardinus	138
ocellatus (1642)	138
pleurophthalmus	138
sanguineus (1643)	138
strigatus (1644)	138
tenuifilis	138
Anthias caballerote	87
macrophthalmus	86
multifasciatus (971)	83
sacer	83
saponaceus	85
vivanus (972)	83
anthias, Labrus	83
antica, Algansea (411)	32
Antimora rostrata	129
viola (1550)	129
antiquorum, Hippocampus	62
antistius var. (846 b)	76
Apeltes quadracus (714)	63
Apherista atricauda (1636)	137
nebulosa (1638)	137
plagiusa (1637)	137
Aphrododeridæ (Family XCVI)	76
Aphredodcrus sayanus (838)	76
Aplodinotus grunniens (1083)	93
Apocope hensbavii	28
nubila	28
ventricosa	28
vulnerata	28
Apodichthys flavidus (1476)	122
fucorum (1477)	122
univittatus (1478)	122
Apogon alutus (1076)	92
americanus	92
imberbis (1073)	92
maculatus (1074)	92
pandionis (1077)	92
retrosella (1075)	92
Apogonichthys	92
Apogonidæ (Family CVII)	92
Apomotis	77
approximans, Polynemus (744)	66
Aprion	87
aprion, Gerres	95
Aprionodon	7
punctatus	8
Apterichthys selachops	52
apua, Epinephelus (988)	84
aquilensis, Lepomis (867)	77
Pomotis	77
arabicus, Chanos	35
aræa, Atherina (726)	65
aræopus, Catostomus (154)	17
Aramaca	133
arara, Serranus	84
aratus, Lutjanus (1016)	87
arcansanum var. (916 b)	80
Archoplites interruptus (844)	76
Archosargus	91
arctica var. (1401 b)	115
arctifrons, Calamus (1061)	91
Citharichthys (1587)	133
Arctozenus	38
arcturus, Salvelinus (528)	44
arcuatus, Pomacanthus	103

	Page.
ardens, Catostomus (166)	18
Notropis (296)	26
ardesiacus, Phoxinus (376)	31
arenatus, Priacanthus	80
arenicola, Fierasfer	126
argentatus, Tetragonopterus (425)	34
argentea, Sphyræna (738)	65
argenteus, Amphistichus (1142)	96
Holconotus (1130)	96
Petromyzon	4
Trachynotus (797)	71
Argentina syrteusium (502)	42
Argentinidæ (Family XLVIII)	42
argentissima, Meda (424)	33
argentiventris, Lutjanus (1006)	87
Mesoprion	87
argentosa, Dionda	21
Argyreus notabilis	28
osculus	28
rubripinnis	27
argyriosus, Pogonichthys	30
Symmetrurus	30
argyritis, Hybognathus (215)	21, 22
Argyropelecus	46
hemigymnus (533)	45
olfersi (534)	45
argyrops, Sparus	91
Argyrosomus	43
argyrosomus, Damalichthys (1149	97
ariommus, Notropis (286)	26
Ariopsis	15
Arius assimilis	15
brandti	15, 16
dasycephalus	15
guatemalensis	15
platypogon	16
seemanni	15
arlingtonia, Gambusia (587)	50
armatus, Leptocottus (1356)	112
Arothron	140
Artedi	4
artedi, Coregonus (513)	43
Artediellus uncinatus (1312)	110
Artedius fenestralis (1307)	110
lateralis (1305)	110
notospilotus (1306)	110
artesiæ, Etheostoma (924)	80
Ascelichthys rhodorus (1301)	109
ascensione, Holocentrum (834)	75
ascensionis, Epinephelus (989)	84
Perca	75
asper, Hexagrammus (1253)	106
Macrurus (1572)	131
aspera, Limanda (1619)	136
Uranidea (1314)	110
Aspidophoroides güntheri (1372)	113
inermis (1370)	113
monopterygius (1369)	113
olriki (1371)	113
asprella, Crystallaria (882)	78
asprellus, Pleurolepis	78
asprigenis, Pœcilichthys	81
aspro, Hadropterus (902)	79
assimilis, Arius	15
asterias, Blennius (1452)	119

	Page.
asterias, Urolophus (81)	11
Asternopteryx gunolliformis	122
Astronesthes niger (493)	42
Astroscopus anoplus (1430)	117, 118
Astyanax	34
Athercsthes stomias (1593)	133
Atherina araea (726)	65
carolina (724)	65
criarcha (723)	65
laticeps	65
stipes (725)	65
velicana	65
Atherinella criarcha	65
Atherinidæ (Family LXXVI)	65
atherinoides, Chriodorus (670)	60
Notropis (308)	27
Atherinops affinis (737)	65
Atherinopsis californiensis (736)	65
atkinsi, Gasterosteus (712)	63
atlanticus, Megalops (434)	34
Rupiscartes (1455)	120
Salarias	120
Atractoscion	95
Atractosteus	13
atrarius, Phoxinus (395)	31
Serranus (958)	82
atricauda, Aphoristia (1636)	137
atrilatus, Zygonectes	50
atrilobatus, Chromis (1194)	102
atripes var. (276 c)	26
Ditrema (1146)	97
atripinnis, Ulocentra	78
atromaculatum var. (885 b)	78
atromaculatus, Semotilus (347)	29
atronasus, Rhinichthys (321)	28
atropurpureus, Anoplarchus (1479)	122
atrovirens, Sebastichthys (1272)	107
attenuatus, Osmerus	42
audens, Menidia (732)	65
auliscus, Siphostoma (685)	61
Aulorhynchidæ (Family LXXIII)	63
Aulorhynchus flavidus (706)	63
Aulostoma maculatum (705)	63
Aulostomidæ (Family LXXII)	63
aurantiacus, Hadropterus (908)	79
aurata, Moniana	25
auratus var. (1078)	92
aureolum, Moxostoma (186)	19, 20
aureus, Ammocœtes (6)	4
Chætodon	103
Pomacanthus (1207)	103
auriculatus, Sebastichthys (1284)	108
auritus, Lepomis (863)	77
aurolineatum, Hæmulon (1042)	89
aurora, Abeona (1134)	96
aurorubens, Rhomboplites (1019)	88
Auxis thazard (765)	68
avocetta, Nemichthys (643)	56
axillaris, Cottus (1342)	111
Pomadasys (1030)	88
Bæostoma	137
Bagropsis	15
bahi di var. (1320 b)	111
Pomatoprion	102
bairdianum, Siphostoma (687)	61

[147] CATALOGUE OF THE FISHES OF NORTH AMARICA.

	Page.
bairdianus, Syngnathus	61
Bairdiella	93
bairdii, Alepocephalus (426)	34
Bathymyzon (12)	4
Gastrostomus (649)	58
Macrurus (1571)	131
Petromyzon	4
bajonado, Calamus (1057)	90
Balistes ciliatus	140
Balistes capistratus (1662)	140
capriscus	140
carolinensis (1659)	140
frenatus	140
mitis	140
polylepis (1661)	140
powelli (1660)	140
vetula (1658)	140
Balistidæ (Family CLIV)	140
balteatus, Pomacanthus	103
Richardsonius (419)	33
Upeneus	93
banana, Gobius (1227)	105
Barathrodemus manatinus (1534)	127
barbaræ, Siphostoma (686)	61
barbatula, Læmonema (1549)	129
barbatum, Echiostoma (491)	42
barbatus, Mullus (1078)	92
Siphagonus (1373)	113
bardus, Pantosteus	17
barratti, Boleosoma	81
bartholomæi, Caranx (784)	70
bascanium, Cæcula	53
Callechelys (616)	52
Bassozetus normalis (1536)	128
Bathymaster	73
signatus (1213)	104
Bathymasteridæ (Family CXIX)	104
Bathymyzon bairdii (12)	4
Bathysaurus agassizii (483)	39, 40
ferox	39
Bathystoma	89
Batrachidæ (Family CXXX)	116
Batrachus tau (1419)	116
pardus (1419 b)	116
bdellium, Petromyzon (8)	4
Bdellostomidæ (Family III)	3
beani, Ammocrypta (878)	77
Caranx	70
Limanda (1620)	136
Ophidion (1527)	126
Poecilichthys	78
Serrivomer (647)	57
bellus, Notropis (300)	27
Belone crassa	59
jonesi	59
stolzmanni	59
Belonidæ (Family LXVIII)	59
bendirei, Uranidea (1319)	111
Benthodesmus elongatus (761)	67
berglax, Macrurus (1568)	131
bernardini, Catostomus (165)	18
Berycidæ (Family XCIV)	74
beryllina, Menidia (733)	65
beryllinus, Cryptotomus (1173)	100
betaurus, Cirrhites	92

	Page.
bicolor, Alganaea (408)	32
Phoxinus (385)	31
bicornis, Cottus	110
Icelus (1308)	110
bifrenatus, Notropis (224)	22
biguttatus, Cochlognathus (222)	22
Hybopsis (328)	28
bilinearis, Merlucius (1565)	131
bilineata, Lepidopsetta (1617)	136
bilineatus, Anisotremus (1037)	89
billingsiana, Cyprinella	24
bilobus, Blepsias (1366)	113
bimaculatus, Chætodon	102
binoculata, Raia (74)	11
birostris, Manta (97)	12
bison var. (148 b)	16
Enophrys (1351)	112
bistrispinus, Bodianus	86
Rhypticus (938)	86
bivittatus, Notropis (295)	26
Platyglossus (1159)	98
blackfordi, Lutjanus	87
blanchardi, Neoclinus (1458)	120
Blennicottus	113
Blenniidæ (Family CXXXVI)	119, 123
blennioides, Diplesion (894)	78
Blennius	120
asterias (1452)	119
brevipinnis	119
carolinus (1453)	119
favosus (1451)	119
fucorum	119
polyactocephalus	122
stearnsi (1450)	119
striatus	119
tripteronotus	121
blennius, Alburnops	24, 26
Etheostoma	78
Notropis (244)	23, 24
Ulocentra (893)	78
Blepharis	70
Blepsias bilobus (1366)	113
cirrhosus (1365)	113
Bodianus	85, 98
bistrispinus	86
bodianus	97
diplotænia (1154)	97
pectoralis (1155)	97
rufus (1153)	97
bodianus, Bodianus	97
Boleichthys	81
elegans	80
warreni	81
boleoides, Uranidea (1329)	111
Boleosoma barratti	81
camurum (888)	78
fusiformis	81
gracile	81
maculatum	78
olmstedi (885)	78
æsopus (885 g)	78
atromaculatum (885 b)	78
effulgens (885 c)	78
maculatum (885 d)	78
mesænum (885 f)	78

	Page.
Boleosoma olmstedi ozarcanum (885 e)	78
susanæ (887)	78
vexillare (886)	78
boleosoma, Gobius (1230)	105
bolli var. (127 c)	15
bombifrons, Lepomis	77
bonaci, Mycteroperca (980)	84
Serranus	84
boops, Myctophum (486)	40
Notropis (243)	24
Scopelus	40
boreale, Etheostoma (932)	80
borealis, Ammocœtes	4
Maurolicus (487)	40
Plagyodus (474)	38
Pœcilichthys	80
Sphyræna (739)	65
Sudis (476)	38
Boreogadus	130
boreum var. (949 c)	81
bosci var. (418 b)	33
Gobiosoma (1243)	106
Menidia	65
Pimelepterus	92
bosquianus, Chasmodes (1439)	119
Bothragonus swani (1377)	114
Bothus maculatus (1576)	132
bouvieri var. (525 b)	44
bovinus, Cyprinodon (547)	47
brachialis, Achirus (1631)	137
brachyacanthus, Amiurus	14
Brachygenys	90
Brachyistius frenatus (1135)	96
rosaceus (1136)	96
Brachyopsis rostratus (1374)	113
verrucosus (1375)	114
xyosternus (1376)	114
brachypoda var. (707 c)	63
brachyptera, Remora (753)	66
brachypterus, Zygonectes	50
brachysomus, Calamus (1058)	90
Brama raji (824)	73
Bramidæ (Family XCI)	73, 104
Branchiostoma lanceolatum (1)	3
Branchiostomidæ (Family I)	3
brandti, Arius	15, 16
Galeichthys (140)	16
brauicki, Pomadasys (1032)	89
brasiliensis, Hemirhamphus	60
Mugil	64
Narcine (78)	11
Scorpæna (1297)	109
breviceps, Larimus (1097)	94
brevipinne, Pristipoma	88
brevipinnis, Blennius	119
Hypsoblennius (1442)	119
Orthopristis (1023)	88
brevirostris, Acipenser (105)	13
Carcharhinus (40)	8
Chasmistes (173)	18
Scomberesox (664)	60
brevispinis var.	107
Sebastichthys (1271)	107
Brevoortia tyrannus (453)	37
patronus (453 b)	37

	Page.
brosme, Brosmius (1554)	130
Brosmius brosme (1554)	130
Brosmophycis ventralis	127
Brotulidæ (Family CXLIV)	125, 126
broussoneti, Umbrina (1104)	94
browni, Stolephorus (460)	37, 38
brunneus, Amiurus (122)	14
Serranus	84
bryoporus, Spratelloides	35
bubalinus, Leuciscus	25
Notropis (253)	25
bubalis, Cottus	111
bubalus, Ictiobus (146)	16
buccata, Ericymba (314)	27
bucco, Moxostoma	19
bufo, Scorpæna	8
bullaris, Semotilus (349)	29
butlerianus, Pœcilichthys	81
Bythites fuscus (1531)	126
caballerote, Anthias	87
caballus var. (785 b)	70
Cæcula bascanium	53
Cæsiosoma californiense (1071)	92
cæsius, Anisotremus (1035)	89
Pomadasys	89
Calamus arctifrons (1061)	91
bajonado (1057)	90
brachysomus (1058)	90
calamus (1056)	90
leucosteus (1059)	91
penna (1061)	91
pennatula	90
proridens (1055)	90
calamus, Calamus (1056)	90
calcarata, Scorpæna	109
californica, Torpedo (77)	11
californicus, Exocœtus (679)	61
Galeus (27)	6, 7
Myliobatis (95)	12
Paralichthys (1595)	133
californiense, Cæsiosoma (1071)	92
Siphostoma (683)	61
californiensis, Atherinopsis (736)	65
Cyprinodon (551)	47
Doryichthys	62
Doryrhamphus (695)	62
Gerres (1127)	95
Scorpis	92
Typhlogobius (1248)	106
Xenistius (1004)	86
callarias, Gadus (1556)	130
Callechelys bascanium (616)	52
scuticaris (614)	52
teres (615)	52, 53
Calliodon	100
calliodon, Liparis (1404)	115
callisema, Notropis (252)	25
callistius, Notropis (266)	25
calliura, Cyprinella	25
calliurus, Ioglossus (1250)	106
callosoma, Novaculichthys	100
calopteryx, Serranus (965)	83
calva, Amia (110)	13
campechanus, Mesoprion	87
Campostoma anomalum (190)	20

CATALOGUE OF THE FISHES OF NORTH AMERICA.

	Page		Page
Campostoma anomalum prolixum (196 b) ...	20	Carcharias taurus	9
formosulum (197)	20	carcharias, Carcharodon (52)	9
ornatum (195)	20	Squalus	8
camura, Cliola	25	Carcharodon carcharias (52)	9
camurum, Boleosoma (888)	78	Careproctus gelatinosus (1395)	115
Etheostoma (920)	80	reinhardti (1396)	115
camurus, Notropis (263)	25	caribæus, Sargus	91
Pœcilichthys	80	Tylosurus (657)	59
canada, Elacate (756)	67	carinatus, Labichthys (644)	56
cauadense, Stizostedion (940)	81	Placopharynx (193)	20
canis, Galeus (26)	6, 7	carminale, Tripterygion (1461)	121
Squalus	6	carminatus, Macrurus (1570)	131
cautharinus, Orthopristis (1024)	88	caruatus, Sebastichthys (1288)	108
Pomadasys	88	carolina, Atherina (724)	65
Cantherhines	140	carolinæ var. (1320 h)	111
Canthogaster	141	carolinensis var. (804 b)	71
capistratus, Balistes (1662)	140	Balistes (1659)	140
Chætodon (1201)	102	carolinus, Blennius (1453)	119
capito, Poromitra (832)	75	Pteraclis (823)	73
capreolus, Epinephelus	84	Trachynotus (796)	71
caprinus, Stenotomus (1062)	91	carpio, Catostomus	19
capriscus, Balistes	140	Cyprinodon (554)	47
caprodes, Percina (699)	79	Ictiobus (147)	16
Carangidæ (Family LXXXV)	69	Carpiodes	16, 17
Caranginæ	70	carringtoni, Agosia (325)	28
Carangoides dorsalis	70	carutta, Johnius	93
Caranx	69	caryi, Hypsurus (1143)	96
amblyrhynchus (782)	70	castanea, Sidera (606)	51
bartholomæi (784)	70	castaneus, Petromyzon (10)	4
beani	70	Catalufa	86
chrysus (785)	70	catalufa, Priacanthus (1000)	86
caballus (785 b)	70	cataphractus var. (713 b)	63
cibi	70	cataractæ, Rhinichthys (320)	27, 28
crinitus (790)	70	catastomus, Phenacobius (317)	27
dorsalis (789)	70	catenatus, Fundulus (569)	49
dumórili	72	Cathorops	15
fallax	70	Catostomidæ (Family XXXI)	16
hippos (787)	70	Catostominæ	18, 19
latus (786)	70	Catostomus aræopus (154)	17
otrynter	70	ardens (166)	18
panamensis	70	bernardini (165)	18
speciosus (788)	70	carpio	19
vinctus (783)	70	catostomus (160)	17
carapinus, Coryphænoides (1574)	131	clarki (155)	17, 18
Carcharhinus	6, 7	commersoni	18
æthalorus (34)	7	congestus	19
brevirostris (40)	8	cypho (168)	18
candatus (37)	8	discobolus (156)	17
cœrulens	8	fecundus (164)	18
fronto (35)	7	guzmaniensis	17
glaucus (32)	7, 8	insignis (169)	18
isodon (42)	8	labiatus (162)	17
lamia (38)	8	latipinnis (157)	17
lamiella (39)	8	longirostris	17
limbatus (41)	8	macrochilus (163)	17
longurio (43)	8	nanomyzon	17
obscurus (33)	7	nebulifer (158)	17
platyodon (36)	7	nigricans	18
terræ-novæ (44)	8	occidentalis (164)	17, 18
Carcharias	6	retropinnis (159)	17
fronto	7	sucetta	19
glaucus	7	tahoensis (161)	17
lamia	8	teres (170)	18
littoralis (49)	9	utawana	18
longurio	8	catostomus, Catostomus (160)	17

	Page.
Catulus	5
catulus	6
var. (125 b)	14
Pimelodus	14
catus, Silurus	14
caudalis, Platyglossus (1160)	98
Pomacentrus (1186)	101
caudata, Lamna	8
caudatus, Carcharhinus (37)	8
Lepidopus (762)	67, 68
Trichiurus	67
caudicula, Conger (637)	55
Caulolatilus chrysops	104
cyanops	104
microps (1216)	104
princeps (1215)	104
Caulolepis longidens (829)	74
caurinus, Mylochilus (352)	30
Sebastichthys (1288)	108
cavalla, Cybium	68
Scomberomorus (769)	68
caelifrons var. (1300 b)	109
Hemitripterus	109
caxis, Lutjanus (1007)	87
Sparus	87
cayuga var. (708 b)	63
Cebedichthys violaceus (1483)	122
Centrarchidæ (Family XCVIII)	76
Centrarchus macropterus (841)	76
Centridermichthys	110
Centriscus scutatus	62
Centropomidæ (Family C)	81
Centropomus medius	82
nigrescens (951)	82
pedimacula (952)	82
robalito (953)	82
undecimalis (950)	81
Centropristis macropoma	82
phœbe	83
radialis	82
Centroscyllium fabricii (18)	5
Centroscymnus cœlolepis (20)	5
centrura, Trygon (85)	12
cepedianum, Dorosoma (455)	36, 37
Cephalacanthus volitans (1393)	115
Cephalocassis	15
Cephalopteridæ (Family XXIV)	12
cephalus, Mugil (715)	64
Ceratias holbölli (1646)	138
Ceratichthys amblops	28
luceus	24–28
micropogon	28
prosthemius	29
sterletus	29
Ceratiidæ (Family CLI)	138
cereostigma, Cyprinella	25
Notropis (260)	25
Cerdale	125
ionthas	126
cerdale, Scytaliscus (1523)	126
Cerdalidæa (Family CXL)	125
cervinum, Moxostoma (192)	20
Cestraciidæ (Family VI)	5
Cestracion francisci (15)	5
philippi	5

	Page.
Cestræus	64
Cetorhinidæ (Family XV)	9
Cetorhinus maximus (53)	9
ceuthœcum, Gobiosoma (1242)	106
Chænobryttus gulosus (846)	75
antistius (846 b)	76
Chænomugil proboscideus (719)	64
Chætodipterus faber (1197)	102
zonatus (1198)	102
Chætodon aureus	103
bimaculatus	102
capistratus (1201)	102
humeralis (1202)	102
maculocinctus (1199)	102
nigrirostris (1203)	102
ocellatus (1200)	102
chætodon, Mesogonistius (852)	76
Chætodontidæ (Family CXVI)	102
chalceus, Orthopristis (1025)	88
chalcogrammus, Pollachius (1562)	130
Chalinura simula (1575)	132
chalybæus, Notropis (282)	26
chalybeius, Hyphalonedrus (503)	42
chamæleonticeps, Lopholatilus (1214)	104
Chanidæ (Family XXXVIII)	35
Chanos arabicus	35
chanos (435)	35
salmoneus	35
chanos, Chanos (435)	35
Mugil	35
Characinidæ (Family XXXIII)	34, 80
Characodon furcidens (555)	48
lateralis	48
Chasmistes brevirostris (173)	18
cujus (175)	18, 19
liorus (172)	18
luxatus (174)	18
Chasmodes bosquianus (1439)	119
quadrifasciatus (1440)	119
saburræ (1441)	119
Chatoëssus signifer	36
Chauliodontidæ (Family LII)	45, 46
Chauliodus sloani (536)	46
Chaunax pictus (1645)	138
chemnitzi, Notacanthus (651)	58
Cheonda	31
chesteri, Phycis (1548)	129
Chiasmodon niger (1437)	119
Chiasmodontidæ (Family CXXXV)	119
chickasavensis, Luxilus	25
chilensis, Sarda (772)	69
chiliticus, Notropis (281)	26
Chilomycterus fuliginosus (1681)	141
geometricus (1680)	141
reticulatus (1682)	141
Chimæra abbreviata	12
affinis (98)	12
colliei (99)	12
plumbea	12
Chimæridæ (Family XXV)	12
chiostictus, Rupiscartes (1454)	120
Salarias	120
Chiridæ (Family CXXII)	106
Chirolophus polyactocephalus (1470)	122
chirurgus, Acanthurus	103

CATALOGUE OF THE FISHES OF NORTH AMERICA.

	Page.
chirus, Xiphister (1480)	122
Chitonotus megacephalus (1310)	110
pugetensis (1311)	110
chloristius, Notropis (269)	25
chlorocephalus, Notropis (280)	26
Chloroscombrus	69
chrysurus (794)	71
orqueta (795)	71
chlorostictus, Sebastichthys (1281)	108
chlorus, Notropis (239)	24
Chologaster agassizii (542)	47
cornutus (541)	47
papillifer (543)	47
Chorinemus altus	72
Chriodorus atherinoides (670)	60
Chromis atrilobatus (1194)	102
enchrysurus (1196)	102
insolatus (1195)	102
punctipinnis (1193)	102
chromis, Pogonias (1084)	93
Chrosomus eos	20
erythrogaster (202)	20
oreas (203)	20
chrosomus, Notropis (283)	26
chrysitis, Diodon	21
chrysochloris, Clupea (442)	36
chrysogaster, Agosia (322)	28
chrysoleucus, Notemigonus (418)	33
Chrysomelas var. (1268 b)	108
chrysops, Caulolatilus	104
Ophichthys (624)	53
Roccus (955)	82
Sparus	91
Stenotomus (1063)	91
chrysoptera, Perca	88
chrysopterum, Hæmulon	89
chrysopterus, Orthopristis (1026)	88
chrysotus, Haplochilus	49
Zygonectes (580)	49
chrysura, Sciæna (1087)	93
chrysurus, Chloroscombrus (794)	71
Ocyurus (1018)	87
chrysus, Caranx (785)	70
chuss, Phycis (1546)	129
cibarius, Ammocœtes (5)	4
cibi, Caranx	70
Cichlidæ (Family CXIII)	101
ciliaris, Holacanthus (1205)	103
ciliatus, Balistes	140
Monacanthus (1063)	140
Sebastichthys (1266)	107
cimbrius, Rhinonemus (1537)	128
cinereus, Gerres (1126)	95
Hadropterus (915)	79
cingulatus, Fundulus	49
Zygonectes	49
cirratum, Ginglymostoma (24)	6
Cirrhisomus	140
Cirrhites betaurus	92
rivulatus (1072)	92
Cirrhitichthys rivulatus	92
Cirrhitidæ (Family CVI)	92
cirrhosus, Blepsias (1365)	113
Cirrostomi	3
Citharichthys	136

	Page.
Citharichthys arctifrons (1587)	133
macrops (1586)	135
microstomus (1589)	133
ocellatus (1579)	133
ovalis (1581)	133
pætulus (1580)	133
panamensis (1582)	133
sordidus (1583)	133
spilopterus (1585)	133
stigmæus (1584)	133
unicornis (1588)	133
Citula	70
clara, Ammocrypta (879)	77
clarki, Catostomus (155)	17, 18
clathratus, Serranus (966)	83
claviformis, Moxostoma	19
claviger, Enophrys	112
Clinidæ	123
Clinocottus	113
Clinostomus	30
Clinus acuminatus	120
evides (1462)	121
xanti	120
zonifer	120
Cliola camura	25
missuriensis	23
nubila	21
topeka	24
urostigma	25
velox	22
vigilax (223)	22
vivax	22
zonata	24
Clupea æstivalis (445)	36
chrysochloris (442)	36
harengus (438)	35
hudsonia	24
humeralis	36
libertatis	37
macrophthalma	36
mediocris (443)	36
mirabilis (439)	35
pensacolæ (449)	36
pseudohispanica (447)	36
sagax (440)	36
sapidissima (446)	36
sardina (447)	36
stolifera (450)	36
thrissa	36
thrissina (448)	36
thryza	36
vernalis (444)	36
Clupeidæ (Family XXXIX)	35
clupeiformis, Coregonus (508)	43
cobitis, Tiarbga (319)	27
coccineus, Lycodes (1515)	124
coccogenis, Notropis (274)	24, 26
Cochlognathus biguttatus (222)	22
ornatus (221)	22
Codoma	24
Cœcula	52
cœlolepis, Centroscymnus (20)	5
cœnosus, Pleuronichthys (1612)	135
cœruleum, Etheostoma (936)	81
cœruleus, Carcharhinus	8

	Page.
cœruleus, Notropis (268)	25
Phoxinus (398)	31
Scarus (1179)	101
Squalus	8
Teuthis (1210)	103
cognata, Uranidea (1321)	111
colias, Scomber (763)	68
Coliscus parietalis	22
collici, Chimæra (99)	12
Colocephali	51
colorado, Lutjanus (1015)	87
comalis, Notropis (240)	24
comifer, Achirus (1632)	137
commersoni, Catostomus	18
complanata, Moniana	24
compressus, Stolephorus (471)	38
concinnus var. (707 b)	63
concolor, Ammocœtes	4
Scomberomorus (766)	68
confertus var. (218 b)	22
Hyborhynchus	22
confluentus, Fundulus (504)	49
conformis, Phoxinus (384)	31
Conger caudicula (637)	55
conger (636)	55
conger, Conger (636)	55
congestum, Moxostoma (188)	19
congestus, Catostomus	19
Congridæ (Family LIX)	52
Congrogadidæ (Family CXLI)	125, 126
Congrogadus	126
coniceps, Murænesox (635)	55
conocephalus, Mylopharodon (353)	30
Conodon nobilis (1020)	88
serrifer (1021)	88
conspersus, Phoxinus (393)	31
constellatus, Sebastichthys (1278)	108
conus, Moxostoma (189)	20
cooperi, Phoxinus (399)	31
copei, Phoxinus (391)	31
copelandi, Cottogaster (895)	78
corallina var. (78 b)	11
Coregonus artedi (513)	43
clupeiformis (508)	43
hoyi (510)	43
kennicotti (506)	43
labradoricus (509)	43
laurettæ (512)	43
merki (511)	43
nelsoni (507)	43
nigripinnis (514)	43
quadrilateralis (505)	43
tullibee (515)	43
williamsoni (504)	43
coregonus, Moxostoma (181)	19
coriaceus, Eleutheractis	85
corinus, Hexanchus (14)	4
Coris	99
cornubica, Lamna (51)	9
cornutus, Chologaster (541)	47
Cyprinus	26
coronatus var. (992)	85
corporalis, Cyprinus	29
coruscans, Sudis	38
Corvina	93

	Page.
Coryphæna dorado	73
equiselis	73
globiceps	73
guttata	73
hippurus (822)	73
lineata	100
nigrescens	82
psittacus	100
punctata	73
sueuri	73
Coryphænidæ (Family XC)	73
Coryphænoides	132
carapinus (1574)	131
rupestris (1573)	131
Coryphopterus	105
Cossyphus puellaris	98
Cottidæ (Family CXXIV)	109
Cottogaster copelandi (895)	78
putnami (896)	78
shumardi (898)	79
uranidea (897)	79
Cottopsis	110
Cottunculus microps (1303)	110
torvus (1304)	110
Cottus æneus (1334)	111
axillaris (1342)	111
bicornis	110
bubalis	111
humilis (1341)	111
jaok	111
labradoricus (1338)	111
niger (1345)	111
octodecimspinosus (1333)	111
platycephalus (1343)	111
polaris	110
polyacanthocephalus (1337)	111
quadricornis (1340)	111
quadrifilis (1346)	111
scorpioides (1335)	111
scorpius (1336)	111
grönlandicus (1336 b)	111
tæniopterus (1339)	111
uncinatus	110
verrucosus (1344)	111
couchi, Moniana	24
couchiana, Pœcilia (592)	50
couesii, Cryptopsaras	139
Couesius dissimilis (343)	29
physignathus (345)	29
plumbeus (344)	29
squamilentus (342)	29
cragini, Amiurus	14
crassa, Belone	59
crassicauda, Phoxinus (394)	31
crassiceps, Plectromus (831)	74
crassilabre, Moxostoma (187)	19
crassus, Alvordius	79
Phoxinus (397)	31
Tylosurus (656)	59
creticula, Zygonectes (578)	49
crebripunctata, Pteroplatea (82)	11
Cremnobates affinis (1467)	121
alticola (1464)	121
fasciatus (1466)	121
integripinnis (1468)	121, 122

CATALOGUE OF THE FISHES OF NORTH AMERICA.

	Page.
Cremnobates marmoratus (1465)	121
nox (1469)	121, 122
crenulare, Myctophum (484)	39
crescentalis, Pomacanthus	103
crinigerum, Siphostoma (694)	62
crinitus, Caranx (790)	70
Cristivomer	44
croicensis, Scarus (1178)	101
crossotus, Etropus (1590)	133
Crotalopsis mordax	53
crumenopthalmus, Trachurops (781)	70
cruoreum, Xiphidium	122
cruoreus, Phoxinus (375)	31
Cryptacanthodes maculatus (1497)	123
Cryptacanthodidæ (Family CXXXVII)	123
Cryptopsaras	138
couesii	139
Cryptotomus beryllinus (1173)	100
roseus	100
ustus (1172)	100
Crystallaria asprella (882)	78
Ctenolabrus adspersus (1150)	97
cubana, Anguilla	55
cubanus, Synodus	39
cubifrons, Malthe	8, 139
cujus, Chasmistes (175)	18, 19
Culius aquideus	105
cumberlandicum var. (923 b)	80
Etheostoma	80
cuningi, Hybopsis (329)	28
curema, Mugil (717)	64
curtus, Stolephorus (465)	38
cuvieri var. (711)	63
cyanellus, Lepomis (853)	77
cyaneus var. (273c)	26
cyanocephalus var. (276d)	26
cyanoguttatus, Heros (1182)	101
cyanolene, Sparisoma (1176)	101
cyanops, Caulolatilus	104
Cybium cavalla	68
petus	68
sara	68
solandri	68
veranyi	68
Cycleptus elongatus (150)	17
Cyclopterichthys stelleri (1408)	116
ventricosus (1407)	116
Cyclopteridæ (Family CXXVIII)	116
Cyclopterus lumpus (1410)	116
cyclopus, Liparis (1405)	115
Cyclothone lusca (537)	46
cylindrostcus	13
cymatotænia, Hadropterus (910)	79
Cynicoglossus pacificus (1628)	136
cynoglossus, Glyptocephalus (1626)	136
Cynoponticus	55
Cynoscion maculatum (1120)	95
nobile (1112)	95
nothum (1115)	95
othonopterum (1116)	95
parvipinne (1117)	95
regale (1113)	95
reticulatum (1119)	95
thalassinum (1114)	95
xanthulum (1118)	95

	Page.
cypho, Catostomus (168)	18
Cyprinella	25
billingsiana	24
callinra	25
cercostigma	25
forbesi	24
gunnisoni	25
notata	25
rubripinna	25
suavis	24
umbrosa	25
cyprinella, Ictiobus (144)	16
Cyprinidæ (Family XXXII)	19, 20, 29
Cyprinodon	48
bovinus (547)	47
californiensis (551)	47
carpio (554)	47
elegans (550)	47
eximius (548)	47
gibbosus	47
latifasciatus (549)	47
macularius (552)	47
mydrus (553)	47
riverendi (546)	47
variegatus (545)	47
gibbosus (545 b)	47
Cyprinodontidæ (Family LIV)	47
Cyprinus americanus	33
cornutus	26
corporalis	29
megalops	26
cyprinus, Ictiobus (149)	17
Cypselurus	61
dactylopterus, Sebastoplus (1293)	109
Dactyloscopus mundus (1424)	117
pectoralis (1425)	117
tridigitatus (1426)	117
Dajaus	64
Dallia pectoralis (60.)	51
Dalliidæ (Family LVIII)	51
dalwigkii, Physiculus	130
Damalichthys argyrosomus (1149)	97
damalis var.	16
dasycephalus, Arius	15
davidsoni, Anisotremus (1036)	89
Monacanthus	140
decagonus, Leptagonus	113
Podothecus (1379)	114
decagrammus, Hexagrammus (1256)	107
Decapterus hypodus	70
macarellus (778)	70
hypodus (778 b)	70
punctatus (777)	69
declivifrons, Glyphidodon (1191)	102
Decodon pnellaris (1156)	98
decoratus, Promicropterus	86
decurreus, Pleuronichthys (1610)	135
dekayi, Isurus (50)	9
delicatissimus, Stolephorus (469)	38
deliciosa, Moniana	23
deliciosus, Notropis (233)	23
Delolepis virgatus (1496)	123
Delothyris pellucidus (1629)	136
delphinus, Pantosteus	17
dentatus, Paralichthys (1596)	134

	Page.
dentatus, Pleuronectes	124
Pseudorhombus	134
Upeneus (1082)	93
dentex, Osmerus (499)	42
depressa, Fistularia (704)	63
Dermatolepis punctatus (995)	85
Diabasis	89
lateralis	90
steindachneri	90
diaphana, Sternoptyx (535)	45, 46
diaphanus, Fundulus (503)	49
Diapterus	95
gracilis	95
harengulus	95
lefroyi	96
dicerans, Enophrys (1352)	112
Dicrolene intronigra (1535)	127
difformis var. (148 d)	16, 17
dilecta, Ancylopsetta (1602)	134
Notosema	134
dilectus, Notropis (309)	27
Dimalacocentrus	100
dimidiata, Algansea (413)	32
dimidiatus, Leucus	32
Dinematichthys	126
marginatus (1532)	127
ventralis (1533)	127
dinemus, Minnilus	27
Diodon hystrix (1678)	141
liturosus (1679)	141
Diodontidæ (Family CLVI)	141
Dionda amara (209)	21
argentosa	21
chrysitis	21
episcopa (210)	21
fluviatilis (208)	21
hæmatura (213)	21
melanops (206)	21
nubila (212)	21
punctifera (207)	21
serena (211)	21
texensis	21
diplæmius, Minnilus	26
Semotilus	26
Diplectrum	82
Diplesion blennioides (894)	78
simoterum	78
Diplodus holbrooki (1067)	91
probatocephalus (1066)	91
rhomboides (1064)	91
unimaculatus (1065)	91
diplotænia, Bodianus (1154)	97
Harpe	97
dipterura, Trygon (86)	12
dipus, Microdesmus (1522)	125, 126
discobolus, Catostomus (156)	17
Discocephali	66
dispar, Zygonectes (577)	49
diaplus, Platyglossus (1163)	99
dissimilis, Ceratichthys (343)	29
Hybopsis (338)	29
Ditrema atripes (1146)	97
furcatum (1147)	97
jacksoni (1145)	97
laterale (1144)	96

	Page.
dolichogaster, Murænoides (1475)	122
dolomici, Micropterus (877)	77
dombeyi, Polistotrema (3)	3
dominus, Notropis (312)	27
dorado, Coryphæna	73
Doratonotus megalepis	99
thalassinus (1167)	99
Dormitator latifrons (1224)	105
maculatus (1223)	105
microphthalmus	105
dormitator, Gobiomorus (1217)	104
Dorosoma cepedianum (455)	36, 37
mexicanum (456)	37
Dorosomidæ (Family XL)	37
dorsalis, Carangoides	70
Caranx (789)	70
Galeus	6
Hypsypops	102
Seriola (807)	72
Umbrina (1103)	94
dorsatus var. (11)	4
Doryichthys californiensis	62
Doryrhamphus californiensis (695)	62
excisus	62
dovii, Anisotremus (1034)	89
Moræna	51
Sidera (608)	51
drummond-hayi, Epinephelus (987)	84
dubius, Ammodytes (749)	66
Fierasfer (1524)	126
ductor, Naucrates (803)	71
dulcis var. (320 b)	27
Rhinichthys	28
dumerili, Caranx	72
Seriola (805)	71, 72
duquesnei var. (185 b)	19
Dussumieria acuta	35
stolifera (436)	35
earlli, Phycis (1545)	129
Echelus	55
Echeneididæ (Family LXXX)	66
Echeneis albescens	66
naucrates (750)	66
Echinorhinus spinosus (16)	5
Echiopsis intertinctus	54
Echiostoma barbatum (491)	42
ectenes, Micropogon (1100)	94
effulgens var. (885 c)	78
eglanteria, Raia (66)	11
egmontis, Myrophis (632)	54
egregius, Phoxinus (381)	31
Elacate canada (756)	67
Elacatidæ (Family LXXXI)	67
Elagatis	69
pinnulatus (810)	72
Elasmobranchii	4
elassochir, Noturus	14
elassodon, Hippoglossoides (1607)	135
Elassoma evergladei (840)	76
zonatum (839)	76
Elassomidæ (Family XCVII)	76
elater, Malthe (1652)	139
elegans, Boleichthys	80
Cyprinodon (550)	47
Gila (358)	30

[155] CATALOGUE OF THE FISHES OF NORTH AMERICA.

	Page.
elegans, Hæmulon	90
Nanostoma	80
Eleotridinæ	105
Eleotris æquidens (1222)	105
amblyopsis (1221)	105
latifrons	105
pisonis (1220)	105
smaragdus	104
Eleuthoractis coriaceus	85
eleutherus, Noturus	14
elongata, Umbrina	94
elongatus var.	88
Benthodesmus (761)	67
Cycleptus (150)	17
Labichthys (645)	56
Lepomis (859)	77
Menticirrus (1106)	94
Ophiodon (1257)	107
Phoxinus (366)	30
Pomadasys (1028)	88
Sebastichthys (1282)	108
Elopidæ (Family XXXVII)	34
Elops saurus (433)	34
Embiotoca	97
Embiotocidæ (Family CXI)	96
Emblemaria nivipes (1450)	120
emiliæ, Opsopœodus (415)	33
emoryi, Gila (363)	30
encæomus, Gobius (1226)	105
Encheylcephali	52
enchrysurus, Chromis (1196)	102
Engraulidæ (Family XLI)	37
Engraulis macrolepidotus	37
perfasciatus	38
Enneacanthus eriarchus (848)	76
gloriosus (850)	76
obesus (849)	76
simulans (851)	76
pinniger (851 b)	76
Enneacentrus fulvus ruber (994)	85
guttatus coronatus (992)	85
tæniops (993)	85
Enophrys bison (1351)	112
claviger	112
diceraus (1352)	112
ensis, Onos (1539)	128
Sphyræna (742)	65
entomelas, Sebastichthys (1268)	107
Eutosphenus	3
Eopsetta	135
eos var. (941 b)	81
Chrosomus	20
Gobiesox (1418)	116
Pœcilichthys	81
Ephippidæ (Family CXV)	102
Ephippus zonatus	102
Epinephelus afer	84
analogus (990)	84
apua (988)	84
ascensionis (989)	84
capreolus	84
drummond-hayi (987)	84
fulvus punctatus	*85
guttatus	84
morio (983)	84

	Page.
Epinephelus nigritus (982)	84
niveatus (986)	84
oxygeneios	83
sellicauda (985)	84
striatus (984)	84
episcopa, Dionda (210)	21
Eques acuminatus (1093)	94
lanceolatus (1094)	94
equisetis, Coryphæna	73
erebennus, Amiurus (128)	15
eriarchs, Atherina (723)	65
Atherinella	65
eriarchus, Enneacanthus (848)	76
Ericosma	79
Ericymba buccata (314)	27
Erimyzon goodei	19
sucetta (176)	19
oblongus (176 b)	19
eriuncea, Raia (63)	11
Erisemus	29
erochrous, Hololepis	81
Erotelis smaragdus (1219)	104
valenciennesi	105
erythrogaster, Chrosomus (202)	20
erythrops, Gobiesox (1417)	116
eschrichti, Oneirodes (1648)	139
esmarki, Lycodes (1511)	124
Esmeralda negra	104
Esocidæ (Family LVI)	50
Esox americanus (597)	50
lineatus	50
lucius (600)	51
lugubrosus	50
nobilior (601)	51
reticulatus (599)	50
salmoneus	50
umbrosus	50
vermiculatus (598)	50
zonatus	49
estor, Phoxinus (368)	30
Etelis	87
Etheostoma artesiæ (924)	80
blennius	78
boreale (932)	80
camurum (920)	80
cœruleum (936)	81
spectabile (936 b)	81
cumberlandicum	80
exile (942)	80
flabellare (923)	80
cumberlandicum (923 b)	80
lineolatum (923 c)	80
fusiforme (941)	81
eos (941 b)	81
histrio	78
inscriptum (919)	80
iowæ (938)	81
jessiæ (937)	81
lepidum (935)	81
luteovinctum (930)	80
lynceum (917)	80
maculatum (921)	80
nevisense	79
nigra	78
parvipinne (931)	80

	Page.
Etheostoma peltatum	79
punctulatum (933)	80
quiescens (940)	81
rufolineatum (922)	80
rupestre (929)	80
sagitta (927)	80
saxatile (928)	80
squamiceps (925)	80
thalassinum (918)	80
tuscumbia (939)	81
variatum	79
virgatum (926)	80
whipplei (934)	81
zonale (916)	80
arcansanum (916 b)	80
Etropus crossotus (1590)	133
Etrumeus micropus	35
teres (437)	35
Eucalia inconstans (708)	63
cayuga (708 b)	63
Eucinostomus lefroyi	96
productus	96
pseudogula	95
Euctenogobius sagittula	105
Eucyclogobius	106
Engomphodus	9
Eulamia	7
lamia	8
longimana	8
Euleptorhamphus longirostris (669)	60
Eumesogrammus præcisus (1484)	122
subbifurcatus (1485)	122
Eumicrotremus spinosus (1409)	116
Eupomotis	77
euryopa, Hudsonius	24
euryorus, Lepomis (871)	77
Eurypharyngidæ (Family LXV)	57, 58
Eurypharynx pelecanoides	58
eurystole, Stolephorus (464)	38
eurystomus, Notropis (264)	25
Euthynnus alliteratus (775)	69
pelamys (776)	69
evansi, Hybognathus	21
Eventognathi	16
evergladei, Elassoma (840)	76
evides, Clinus (1462)	121
Hadropterus (905)	79
evolans, Halocypselus (672)	60
Prionotus (1390)	115
Trigla	115
exasperatus, Rhinobatus (61)	10
excisus, Doryrhamphus	62
exiguus, Stolephorus (467)	38
exile, Etheostoma (942)	81
exilicauda, Lavinia (201)	20
exiliens, Exocœtus (673)	61
exilis, Hippoglossoides (1608)	135
Noturus (117)	14
Tylosurus (661)	59
eximius, Cyprinodon (548)	47
Exocœtus affinis	61
californicus (679)	61
exiliens (673)	61
furcatus (678)	61
gibbifrons (680)	61

	Page.
Exocœtus heterurus (677)	61
hillianus	60
melanurus	61
mesogaster	60, 61
obtusirostris	60
roberti	61
rondeleti (674)	61
vinciguerræ (675)	61
volador	61
volitans (676)	61
Exoglossum maxillingua (220)	22
extensus, Fundulus (562)	49
faber, Chætodipterus (1197)	102
fabricii, Centroscyllium (18)	5
Macrurus	131
falcata, Mycteroperca (978)	84
Seriola	72
falcatus, Labrus	97
Lachnolæmus	97
fallax, Caranx	70
Fario	44
fasciata, Seriola (808)	72
fasciatum, Pristipoma	88
fasciatus, Achirus	137
Cremnobates (1406)	121
Hadropterus (906)	79
Hemirhamphus	60
Larimus (1096)	94
Murænoides (1472)	122
Prionodes	83
Trachurus	72
Trachynotus (802)	71
favosus, Blennius (1451)	119
fecundus, Catostomus (167)	18
feliceps, Galeichthys	15
felis, Galeichthys (138)	16
fenestralis, Artedius (1307)	110
ferox, Bathysaurus	39
Plagyodus (472)	38
Stomias (480)	41
ferruginea, Limanda (1618)	136
Fierasfer arenicola	126
dubius (1524)	126
Fierasferidæ (Family CXLII)	126
fimbria, Anoplopoma (1261)	107
Fistularia depressa (704)	63
serrata (703)	63
tabaccaria (702)	63
Fistulariidæ (Family LXXI)	63
flabellare, Etheostoma (923)	80
flagellum, Saccopharynx	57
flammeus, Phoxinus (403)	31
flavescens, Sparisoma (1177)	101
flavidus, Apodichthys (1476)	122
Aulorhynchus (706)	63
Sebastichthys (1264)	107
flaviguttatum, Hæmulon (1041)	89
flavilatus, Pomacentrus (1188)	102
flavipinnis, Hybognathus	21
flavolineatum, Hæmulon (1045)	90
flavovittatus, Mulloides	93
Upeneus	93
flavus, Noturus (110)	14
florealis, Platyglossus	98
floridæ, Jordanella (544)	47

[157] CATALOGUE OF THE FISHES OF NORTH AMERICA.

	Page.
floridæ, Siphostoma (689)	62
floridanus, Phycis (1544)	129
floripinnis, Zygonectes (573)	49
fluviatilis, Dionda (208)	21
Hudsonius	24
fodiator, Tylosurus (655)	59
fœtens, Synodus (477)	39
fonticola, Alvarius (946)	81
Micropercn	81
fontinalis, Salvelinus (530)	44
forbesi, Cyprinella	24
formosa, Algansea	32
Heterandria (593)	50
Uranidea (1331)	111
formosulum, Campostoma (197)	20
formosus, Notropis (251)	25
Serranus (961)	82
forsteri, Sphyræna	65
francisci, Cestracion (15)	5
franklini, Uranidea (1330)	111
fremebundum, Hæmulon (1049)	90
freminvillei, Myliobatis (94)	12
frenatus, Balistes	140
Brachyistius (1135)	96
fretensis, Notropis (230)	23
frigida, Moniana	24
frontalis var. (273 b)	26
fronto, Carcharhinus (35)	7
Carcharias	7
fucorum, Apodichthys (1477)	122
Blennius	119
fulgida, Meda (423)	33
fuliginosus, Chilomycterus (1681)	141
fulvomaculatum, Pristipoma	88
fulvus var.	85
Enneacentrus (994)	85
Epinephelus	85
Labrus	85
Physiculus (1551)	130
funduloides, Phoxinus (369)	30
Fundulus adinia (565)	49
catenatus (569)	49
cingulatus	49
confluentus (564)	49
diaphanus (563)	49
extensus (562)	49
heteroclitus (566)	49
grandis (566 b)	49
majalis (557)	48
menona	49
nigrofasciatus	49
ocellaris (567)	49
parvipinnis (559)	48
seminolis (561)	49
similis (558)	48
stellifer (570)	49
swampina	48
vinctus (568)	49
xenicus	48
zebrinus (560)	48
funebris, Gymnothorax	52
Muræna	52
Noturus (114)	14
Sidera (610)	52
furcatum, Ditrema (1147)	97

	Page.
furcatus, Exocœtus (678)	61
Ictalurus (135)	15
furcidens, Characodon (555)	48
furcifer, Paranthias (973)	83
furva, Perca	82
furvus, Serranus (959)	82
fuscum, Siphostoma (692)	62
fuscus, Bythites (1531)	126
fusiforme, Etheostoma (941)	81
fusiformis, Boleosoma	81
Gadidæ (Family CXLV)	126, 128
Gadus callarias (1556)	130
gracilis	130
navaga	130
ogac (1557)	130
gaimardianus, Mugil (716)	64
gairdneri, Salmo (524)	44
galacturus, Notropis (262)	25
galeatus, Gymnacanthus (1349)	112
Galeichthys brandti (140)	16
feliceps	15
felis (138)	16
guatemalensis (136)	15
platypogon (139)	16
seemanni (137)	15
Galeocerdo maculatus (31)	7
tigrinus	7
Galeorhinidæ (Family X)	6
Galeorhinus	6
zyopterus (30)	7
Galeus	9
californicus (27)	6, 7
canis (26)	6, 7
dorsalis	6
lunulatus (25)	6
maculatus	7
mustelus	6
galeus, Squalus	6
galtiæ, Phoxinus (374)	31
Squalius	31
Gambusia affinis (588)	50
arlingtonia (587)	50
holbrooki	50
humilis (586)	50
nobilis (589)	50
patruelis (585)	50
senilis (590)	50
gardoneus, Notemigonus (417)	33
garmani, Lepomis (865)	77
Notropis (256)	25
Gasterosteidæ (Family LXXIV)	63
Gasterosteus aculeatus (713)	63
cataphractus (713 b)	63
atkinsi (712)	63
(cuvieri?) wheatlandi (711 b)	63
(gymnurus?) cuvieri (711)	63
microcephalus (710)	63
williamsoni (709)	63
Gastrostomus bairdii (649)	58
gelatinosum, Melanostigma (1521)	125
gelatinosus, Careproctus (1395)	115
gelidus, Hybopsis (341)	29
geminatus, Hypleurochilus (1449)	119
gemma, Hypoplectrus (970)	38
generosus, Pantosteus (152)	17

	Page.
gentilis, Hypsoblennius (1443)	119
Genyonemus lineatus (1008)	94
Genypterus omostigma	120
geometricus, Chilomycterus (1680)	141
georgianus, Scorpis	92
Gerres aprion	95
californiensis (1127)	95
cinereus (1126)	95
gracilis (1129)	95, 96
gula (1128)	95
homonymus	95
jonesi (1130)	95
lefroyi (1131)	95
lineatus (1123)	95
olisthostoma (1124)	95, 96
peruvianus (1125)	95
plumieri (1122)	95
rhombeus	96
Gerridæ (Family CX)	95
gibba, Liparis (1300)	115
gibbifrons, Exocœtus (680)	61
Gibbonsia	121
gibbosa	90
Moniana	24
Perca	90
gibbosum, Hæmulon (1052)	90
gibbosus var. (545 b)	47
Cyprinodon	47
Lepomis (875)	77
Leuciscus	31
gigas, Stereolepis (975)	83
Gila affinis (361)	30
elegans (358)	30
emorii (363)	30
gracilis (362)	30
grahami (360)	30
nacrea (364)	30
robusta (359)	30
seminuda (366) (365)	30
gilberti, Hypsoblennius (1444)	119
Notropis (235)	23
Gillichthys mirabilis (1237)	106
Ginglymodi	13
Ginglymostoma	8
cirratum (24)	6
Girardinus	50
Girella	92
nigricans (1068)	91
glaber, Pleuronectes (1623)	136
glaciale, Myctophum	40
glacialis, Pleuronectes (1624)	136
gladius, Istiophorus	67
Tylosurus	59
Xiphias (757)	67
Glaniostomi	13
glaucofrænum, Gobius (1234)	105
glaucostigma, Rhinobatus (59)	10
glaucus, Carcharhinus (32)	7, 8
Carcharias	7
Trachynotus (801)	71
globiceps, Coryphæna	73
Oligocottus (1364)	113
gloriosus, Enneacanthus (850)	76
Glossamia	92
glutinosa, Myxine (2)	3

	Page.
Glyphidodon declivifrons (1191)	102
saxatilis (1192)	102
troscheli (1192 b)	102
troscheli	102
Glyptocephalus cynoglossus (1626)	136
zachirus (1627)	136
Gnathanodon	70
Gnathypops maxillosus (1433)	118
mystacinus (1432)	118
rhomaleus (1431)	118
Gobiesocidæ (Family CXXIX)	116
Gobiesox adustus (1415)	116
eos (1418)	116
erythrops (1417)	116
mæandricus (1411)	116
rhessodon (1414)	116
strumosus (1412)	116
virgatulus (1413)	116
zebra (1416)	116
Gobiidæ (Family CXXI)	104
Gobio æstivalis	29
plumbeus	29
gobioides, Hypsicometes (1564)	131
Uranidea (1328)	111
Gobiomorus dormitator (1217)	104
lateralis (1218)	104
Gobionellus	105
oceanicus (1235)	104, 106
stigmaticus (1236)	106
Gobiosoma bosci (1243)	106
enthœcum (1242)	106
histrio (1244)	106
ios (1247)	106
longipinne (1246)	106
zosterurum (1245)	106
Gobius banana (1227)	105
boleosoma (1230)	105
encæomus (1226)	105
glaucofrænum (1234)	105
lyricus (1225)	105
nicholsi (1233)	105
sagittula (1229)	105
soporator (1228)	105
stigmaturus (1231)	105
würdemanni (1232)	105
goodei, Erimyzon	19
Halosaurus	41
Lucania (584)	49
Ptilichthys (650)	58
Spinivomer (646)	57
gorbuscha, Oncorhynchus (518)	43
goreensis, Trachynotus	71
gouani, Lepidopus	67
gracile, Boleosoma	81
gracilirostris, Histiophorus	67
gracilis, Diapterus	95
Gadus	130
Gerres (1129)	95, 96
Gila (362)	30
Hybopsis	28
Moniana	24
Phoxinus (383)	31
Platygobio (346)	29
Uranidea (1327)	111
graëllsi, Ophidium	126

CATALOGUE OF THE FISHES OF NORTH AMERICA.

	Page.
grahami, Gila (360)	30
grandicornis, Scorpæna (1296)	109
grandis var. (566 b)	49
grandisquamis, Platyglossus	08
Upeneus (1081)	93
granulata, Raia (69)	11
grayi var.	16
griseolineatum, Siphostoma (684)	61
griseum var. (949 b)	81
griseus, Labrus	87
Lutjanus (1009)	87
grœnlandicum, Microstoma (494)	42
grœnlandicus, Himantolophus (1649)	139
Grouias nigrilabris (121)	14
grönlandicus var. (1336 b)	111
gronovii, Nomeus (815)	72
grunniens, Aplodinotus (1083)	93
guacamaia, Scarus (1180)	101
gnaguanche, Sphyræna (740)	65
guasa, Promicrops	84
guatemalensis, Arius	15
Galeichthys (136)	15
gula, Gerres (1128)	95
Phoxinus (379)	31
gulosa, Uranides (1317)	111
gulosus, Chænobryttus (846)	76, 77
Lepidogobius (1240)	106
gunelliformis, Murænoides	122
gunnellus, Murænoides (1471)	122
gunuisoni, Cyprinella	25
guntheri, Aspidophoroides (1372)	113
Hoplopagrus (1005)	86, 87
guttata, Coryphæna	73
Scorpæna (1294)	109
guttatus, Euneacentrus (962)	85
Epinephelus	84
Lamprie (821)	73
Lutjanus (1011)	87
Percopsis (532)	44
Upsilonphorus (1429)	117
guttifer, Ophichthys (622)	53
guttulata, Hypsopsetta (1613)	135
guzmaniensis, Catostomus	17
Pantosteus (153)	17
Gymnacanthus galeatus (1349)	112
pistilliger (1348)	112
tricuspis (1347)	112
Gymnelis stigma	125
viridis (1519)	125
Gymnomuræna nectura	51
Gymnothorax afer	52
funebris	52
gymnothorax, Thymallus	43
gyrans, Querimana (721)	64
gyrinus, Noturus (111)	14
Gyropleurodus	5
Hadropterus aspro (902)	79
aurantiacus (908)	79
cinereus (915)	79
cymatotænia (910)	79
evides (905)	79
fasciatus (906)	79
macrocephalus (900)	79
maculatus	79
nianguæ (911)	79

	Page.
Hadropterus nigrofasciatus (907)	79
ouachitæ (903)	79
peltatus (904)	79
phoxocephalus (901)	79
scierus (913)	79
squamatus (909)	79
tessellatus (914)	79
variatus (912)	79
hæmatura, Dionda (213)	21
Hæmulon acutum (1051)	90
aurolineatum (1042)	89
chrysopterum	89
elegans	90
flavignttatum (1041)	89
flavolineatum (1045)	90
fremebundum (1049)	90
gibbosum (1052)	90
jeniguano	89
maculicauda (1040)	89
plumieri (1046)	90
rimator (1043)	89
sciurus (1047)	90
scudderi (1050)	90
sexfasciatum (1053)	90
steindachneri (1048)	90
tæniatum (1044)	90
Hæmulopsis	88
Halecomorphi	13
Halieutæa senticosa (1654)	139
Halieutichthys aculeatus (1653)	139
Haliperca phœbe	83
halleri, Urolophus (80)	11
Halocypselus evolans (672)	60
Haloporphyrus rostratus	129
viola	129
Halosauridæ (Family XLVI)	41
Halosaurus goodei	41
macrochir (488)	41
oweni	41
hamatus, Icelus	110
Haplochilus chrysotus	49
melanops	50
Haplomi	47
Harengula sardina	36
harengulus, Diapterus	95
harengus, Clupea (438)	35
Myxus	64
Querimana (720)	64
harfordi, Ptychochilus (356)	30
Harpe	97, 98
diplotænia	97
pectoralis	97
Harpodon	39
hastata, Trygon (86)	12
hayi, Hybognathus (217)	22
Hemiarius	15
Hemibranchii	62
Hemicaranx	70
hemigymnus, Argyropelecus (533)	45
Hemilepidotus hemilepidotus (1359)	112
jordani (1358)	112
spinosus (1357)	112
hemilepidotus, Hemilepidotus (1359)	112
Hemirhamphus brasiliensis	90
fasciatus	60

	Page.
Hemirhamphus picarti	60
pleei (668)	60
povyi	60
richardi	60
roberti (666)	60
rosæ (667)	60
unifasciatus (665)	60
Hemirhombus ovalis	133
pætulus	133
Hemistoma	100
Hemitremia vittata	22
Hemitripterus americanus (1300)	109
cavifrons (1300 b)	109
cavifrons	109
henlei, Triacis (29)	7
henshalli, Zygonectes (572)	49
henshavii, Apocope	28
henshawi, var. (525 d)	44
hentzi, Hysoblennius	119
Isesthes	119
hepatus, Teuthis (1208)	103
heptagonus, Hippocampus	62
Heptranchias maculatus (13)	4
heraldi, Tetrodon	141
Heros cyanoguttatus (1182)	101
pavonaceus (1183)	101
heros, Lepomis (870)	77
Heterandria formosa (593)	50
occidentalis (594)	50
ommata (595)	50
heteroclitus, Fundulus (566)	49
Heterodon	5
heterodon, Notropis (236)	22
Heterodontus	5
Heterosomata	132
Heterostichus	120
rostratus (1463)	121
heterurus, Exocœtus (677)	61
Hexagrammus asper (1253)	106
decagrammus (1256)	107
ordinatus (1252)	106
scaber (1254)	107
superciliosus (1255)	107
Hexanchus corinus (14)	4
Hexanematichthys	15
hians, Tylosurus (654)	59
Hiatula ouitis (1151)	97
hillianus, Exocœtus	60
Himantolophus grœnlandicus (1649)	139
reinhardti (1650)	139
Hippocampidæ	62
Hippocampus antiquorum	62
heptagonus	62
hippocampus	62
hudsonius (698)	62
ingens (696)	62
punctulatus (697)	62
stylifer (699)	62
zosteræ (700)	62
hippocampus, Hippocampus	62
Hippoglossina macrops (1604)	135
Hippoglossoides elassodon (1607)	135
exilis (1608)	135
jordani (1605)	135
platessoides (1606)	135

	Page.
hippoglossoides, Reinhardtius (1502)	133
Hippoglossus	135
hippoglossus (1591)	133
hippoglossus, Hippoglossus (1591)	133
hippos, Caranx (787)	70
hippurus, Coryphæna (822)	73
hirudo, Petromyzon (9)	4
hirundo, Liocottus (1353)	112
hispidus, Monacanthus (1664)	140
Histiobranchus infernalis (641)	56
Histiophorus americanus	67
ancipitirostris	67
gracilirostris	67
histrio, Etheostoma	78
Gobiosoma (1244)	106
Pterophrynoides (1640)	138
Ulocentra (892)	78
Holacanthus ciliaris (1205)	103
striqatus (1204)	103
tricolor	103
holbolli, Ceratias (1646)	138
holbrooki, Diplodus (1067)	91
Gambusia	50
Lepomis (873)	77
Ophidion (1526)	126
Holconotus agassizii (1140)	91
analis (1138)	96
argenteus (1139)	96
rhodoterus (1141)	96
Holocentridæ (Family XCV)	75
Holocentrum ascensione (834)	75
matejuelo	75
pentacanthum	75
suborbitale (835)	75
Holocephali	12
Hololepis erochrous	81
homonymus, Gerres	95
Hoplopagrus güntheri (1005)	86, 87
Hoplostethus mediterraneus (833)	75
hoyi, Coregonus (510)	43
Uranides (1332)	111
hudsonia, Clupea	24
Hudsonius curyopa	24
fluviatilis	24
hudsonius, Hippocampus (698)	62
Notropis (246)	24
humboldti, Phoxinus (373)	30
humeralis, Chætodon (1202)	102
Clupea	36
Platyglossus	98
humilis, Cottus (1341)	111
Gambusia (586)	50
Lepomis (868)	77
Hybognathus argyritis (215)	21, 22
evansi	21
flavipinnis	21
hayi (217)	22
meeki (214)	21
nigrotæniatus	21
nuchalis (216)	21
placita (216 b)	21
regia (216 c)	22
osmerinus	21
placita	21
regius	21

CATALOGUE OF THE FISHES OF NORTH AMERICA.

	Page.
Hybopsis	24
æstivalis (340)	29
amblops (331)	29
rubrifrons (331 b)	29
biguttatus (328)	28
cumingi (329)	28
dissimilis (333)	29
gelidus (341)	29
gracilis	28
hyostomus (337)	29
hypsinotus (332)	29
labrosus (336)	29
longiceps	23
marconis (339)	29
monachus (334)	29
montanus (338)	29
plumbeolus	26
storerianus (330)	28
tuditanus	22
volucellus	23
zanemus (335)	29
Hyborhynchus confertus	22
superciliosus	22
Hydrargyra	48
Hydrolagus	12
Hydrophlox	26
hydrophlox, Phoxinus (370)	30
Hyodon alosoides (430)	34
selenops (432)	34
tergisus (431)	34
Hyodontidæ (Family XXXVI)	34
Hyoprorus	54
hyostomus, Hybopsis (337)	29
Nocomis	29
Hypargyrus tuditanus	22
Hypentelium nigricans (171)	18
Hyperchoristus tanneri (490)	41, 42
Hyperotreta	3
Hyperprosopon	96
Hyphalonedrus chalybeius (503)	42
Hypleurochilus geminatus (1449)	119
multifilis (1448)	119
Hypocritichthys	96
hypodus var. (778 b)	70
Decapterus	70
Hypomesus olidus (501)	42
pretiosus (500)	42
Hypoplectrus gemma (970)	83
nigricans (969)	83
Hypoprion	7, 8
hypselopterus, Notropis (272)	25
Hypsicometes gobioides (1564)	131
Hypsilepis iris	24
hypsinotus, Hybopsis (332)	29
Hypsoblennius brevipinnis (1442)	119
gentilis (1443)	119
gilberti (1444)	119
hentzi	110
jenkinsi (1446)	119
punctatus (1445)	119
scrutator (1447)	119
Hypsopsetta guttulata (1613)	135
Hypsurus caryi (1143)	96
Hypsypops dorsalis	102
Hysterocarpus traski (1132)	96

	Page.
hystrix, Diodon (1678)	141
Icelinus quadriseriatus (1309)	110
Icelus bicornis (1308)	110
hamatus	110
Ichthyapus acutirostris	52
selachops (612)	52
Ichthyomyzon	4
Icichthys	104
lockingtoni (826)	73
Icistia, Sciæna (1088)	93
Icosteidæ (Family XCII)	73, 104
Icosteus	104
ænigmaticus (825)	73
Ictalurus furcatus (135)	15
punctatus (134)	15
Ictiobus bubalus (146)	16
carpio (147)	16
cyprinella (144)	16
cyprinus (149)	17
urus (145)	16
velifer (148)	16, 17
bison (148 b)	16
difformis (148 d)	17
tumidus (148 c)	16
illecebrosus, Alburnops	23
Notropis (229)	23, 24
imberbe, Peristedinm (1583)	114
imberbis, Apogon (1073)	92
immaculatus var. (530 b)	44
inconstans, Eucalia (708)	63
inermis var. (72 b)	11
Aspidophoroides (1370)	113
Lutjanus (1017)	87
infernalis, Histiobranchus (641)	56
Muræna	52
ingens, Hippocampus (696)	62
Iniistius	100
mundicorpus	100
Iniomi	45
Iuopsetta	136
inornatus, Raia (72)	11
inornatus, Microlepidotus	88
Orthopristis (1022)	88
Pomadasys	88
inscripta, Solea	137
inscriptum, Etheostoma (919)	80
inscriptus, Achirus (1634)	137
insignis, Catostomus (169)	18
insolatus, Chromis (1195)	102
integripinnis, Cremnobates (1468)	121, 122
intermedius, Phoxinus (389)	31
Saurus	39
Synodus	39
interruptus, Anisotremus (1036)	89
Archoplites (844)	76
Roccus (956)	82
intertinctus, Echiopsis	54
Ophichthys (627)	53
Ophisurus	54
intronigra, Dicrolene (1535)	127
inurus, Zygonectes	50
Ioa vigilis (884)	78
vitrea (883)	78
Ioglossus calliurus (1250)	106

	Page.
ionthas, Cerdale	126
Hypsoblennius (1446)	119
iox, Gobiosoma (1247)	106
iowæ, Etheostoma (938)	81
iridens var. (524 b)	44
Salmo	44
iris, Hypsilepis	24
ischanus, Stolephorus (462)	38
ischyra, Isopsetta (1616)	136
ischyrus, Lepomis (856)	77
Parophrys	136
Isesthes hentzi	119
punctatus	119
Isodon, Carcharhinus (42)	8
Isogomphodon	8
isolepis, Isopsetta (1615)	135, 136
Isopsetta ischyra (1616)	136
isolepis (1615)	135, 136
Isospondyli	34, 45
Istiophorus americanus (759)	67
gladius	67
Isuropsis	9
Isurus dekayi (50)	9
itaiara, Promicrops (076)	64
Serranus	84
jacksoni, Ditrema (1145)	97
jacobi, Sciæna (1089)	93
jacobus, Myripristis	75
jaok, Cottus	111
japonicus, Trichodon (1423)	117
jarrovii, Lepidomeda (422)	33
jejunus, Notropis (288)	26
jemezanus, Alburnellus	27
jeniguano, Hæmulon	89
jessiæ, Etheostoma (937)	81
Poecilichthys	81
jocu, Lutjanus (1008)	87
Johnius carutta	93
saturnus (1092)	93
saxatilis	94
jonesi, Belone	59
Gerres (1130)	95
Jordanella floridæ (544)	47
jordani, Hemilepidotus (1358)	112
Hippoglossoides (1605)	135
josephi, Ophidium	126
jugalis, Moniana	24
Julis lucasana	99
maculipinna	99
purpureus	99
julis, Labrus	99
kennedyi, Trachynotus (799)	71
kennerlyi, Moxostoma	19
kennicotti, Coregonus (506)	43
keta, Oncorhynchus (519)	44
kisutch, Oncorhynchus (521)	44
kuhli, Sebastes	108
kumlieni var. (1320 c)	111
Kyphosus analogus (1070)	92
sectatrix (1069)	92
labiatus, Catostomus (162)	17
Labichthys carinatus (644)	56
elongatus (645)	56
Labidesthes sicculus (728)	65
labradoricus, Coregonus (509)	431

	Page.
labradoricus, Cottus (1338)	111
Labridæ (Family CXII)	97
Labrosomus nuchipinnis (1459)	120
xanti (1459 b)	120
xanti	120
zonifer (1460)	120
labrosus, Hybopsis (336)	29
Labrus anthias	83
falcatus	97
fulvus	85
griseus	87
julis	99
maximus	97
radiatus	98
rufus	97
lacera, Quassilabia (194)	20
lacertosus, Notropis (285)	26
Lachnolæmus falcatus	97
maximus (1152)	97
suillus	97
laciniata, Menidia (729)	65
Lactophrys	139
Læmonema barbatula (1540)	129
lætabilis, Moniana	24
lævigatus, Lagocephalus (1669)	140
lævis, Raia (75)	11
Lagocephalus lævigatus (1669)	140
Lagodon	91
lalandi var. (805 b)	71
Seriola	71, 72
lampetræformis, Leptoblennius (1494)	123
lamia, Carcharhinus (38)	8
Carcharias	8
Eulamia	8
lamiella, Carcharhinus (39)	8
Lamna caudata	8
cornubica (51)	9
Lamnidæ (Family XIV)	9
Lampetra	4
Lampridiæ (Family LXXXIX)	73
Lampris guttatus (821)	73
lanceolata, Sciæna (1086)	93
lanceolatum, Branchiostoma (1)	3
lanceolatus, Eques (1094)	94
Larimus breviceps (1097)	94
fasciatus (1096)	94
laterale, Ditrema (1144)	96
lateralis, Alvarius (943)	81
Artedius (1305)	110
Characodon	48
Diabasis	90
Gobiomorus (1218)	104
Philypnus	104
Richardsonius (420)	33
laticeps, Aëtobatis	12
Atherina	65
Stoasodon (93)	12
latifasciatus, Cyprinodon (549)	47
latifrons, Anarrhichas (1500)	123
Dormitator (1224)	105
Eleotris	105
Notarus (115)	14
Latilidæ	104
Latilus	104
atipinna, Mollienesia (591)	50

[163] CATALOGUE OF THE FISHES OF NORTH AMERICA.

	Page.
latipinnis, Catostomus (157)	17
Zaniolepis (1258)	107
latus, Caraux (786)	70
laurettæ, Coregonus (512)	43
Lavinia exilicauda (201)	20
lefroyi, Diapterus	96
Eucinostomus	90
Gerres (1131)	95
Leirus perciformis (820)	73
lemmoni, Squalius	31
lentiginosus, Rhinobatus (60)	10
leonina, Moniana	24
leoninus, Notropis (248)	24
leopardinus, Antennarius	138
Platophrys (1577)	132
Rhomboidichthys	132
Lepidogobius gulosus (1240)	106
lepidus (1238)	106
newberryi (1239)	106
thalassinus (1241)	106
Lepidomeda jarrovii (422)	33
vittata (421)	33
Lepidopsetta bilineata (1617)	136
Lepidopus caudatus (762)	67, 68
gonani	67
Lepidosteidæ (Family XXVIII)	13
Lepidosteus osseus (107)	13
platystomus (108)	13
spatula	13
tristœchus (109)	13
lepidum, Etheostoma (935)	81
lepidus, Lepidogobius (1238)	106
Notropis (254)	25
Lepomis albulus (872)	77
aquilensis (867)	77
auritus (863)	77
bombifrons	77
cyanellus (853)	77
elongatus (850)	77
euryorus (871)	77
garmani (865)	77
gibbosus (875)	77
heros (870)	77
holbrooki (873)	77
humilis (868)	77
ischyrus (856)	77
lirus	77
macrochirus (857)	77
marginatus (866)	77
megalotis (864)	77
miniatus (862)	77
murinus (860)	77
mystacalis (858)	77
notatus (874)	77
pallidus (869)	77
phenax (855)	77
punctatus (861)	77
symmetricus (854)	77
leptacanthus, Noturus (112)	14
Leptagonus	114
decagonus	113
Leptarius	15
Leptoblennius lampetræformis (1494)	123
nubilus (1492)	123
serpentinus (1493)	123

	Page.
Leptocardii	3
Leptocephalus morrisi	55
Leptoclinus maculatus (1488)	123
Leptocottus armatus (1356)	112
Leptophidium profundorum (1530)	126
Leptops olivaris (120)	14
leptorhynchum, Siphostoma (688)	62
Leptoscopidæ (Family CXXXII)	117
lepturus, Anarrhichas (1501)	123
Trichiurus (760)	67
Letharchus velifer (613)	52
lethostigma, Paralichthys (1607)	134
leucichthys, Stenodus	43
leuciodus, Notropis (289)	26
Leuciscus bubalinus	25
gibbosus	31
lutrensis	24
leuciscus, Pomadasys (1027)	88
leucopus, Rhampoberyx	76
Leucos	32
leucosteus, Calamus (1059)	91
leucostictus, Pomacentrus (1185)	101
leucotænia, Pholidichthys	123
Leucus dimidiatus	32
olivaceus	52
Leuresthes tenuis (727)	65
libertate, Opisthonema (452)	37
libertatis, Clupea	37
Meletta	37
Limanda aspera (1619)	136
beani (1620)	136
ferruginea (1618)	136
limbatus, Carcharhinus (41)	8
limi, Umbra (596)	50
lineata, Coryphæna	100
Sciæna	82
Trigla	115
lineatus var.	115
Achirus	137
Esox	50
Genyonemus (1098)	94
Gerres (1123)	95
Phoxinus (382)	31
Phtheirichthys (751)	66
Pleuronectes	137
Roccus	82
Tetrodon	140
Xyrichthys	100
Zygonectes (574)	49
lineolata, Mollienesia	50
lineolatum var. (923 c)	80
lineopinnis, Murana	52
Liocottus hirundo (1353)	112
liolepis, Xystreurys (1603)	135
Lioperca	85
liorus, Chasmistes (172)	18
Liostomus xanthurus (1095)	94
Liparidæ (Family CXXVII)	115
liparina, Amitra	115
Monomitra (1394)	115
Liparis calliodon (1404)	115
cyclopus (1405)	115
gibba (1399)	115
liparis (1401)	115
arctica (1401 b)	115

	Page.
Liparis major (1397)	115
montaguei (1403)	115
mucosa (1406)	115
pulchella (1398)	115
ranula (1402)	115
tunicata (1400)	115
Liparis, Liparis (1401)	115
lirus, Lepomis	77
Notropis (302)	27
littoralis, Carcharias (49)	9
Menticirrus (1105)	94
liturosus, Diodon (1679)	141
lividus var. (127 b)	15
liza, Mugil	64
Lobotes surinamensis (1002)	80
Lobotidæ (Family CIV)	86
lockingtoni, Icichthys (826)	73
lonchura, Opisthognathus (1435)	118
longa, Trygon (88)	12
longiceps var. (233c)	23
Hybopsis	23
longicollis, Myrophis	54
longidens, Caulolepis (829)	74
longimana, Eulamia	8
longimanus, Squalus	8
longipinne, Gobiosoma (1246)	106
longirostris, Catostomus	17
Euleptorhamphus (669)	60
Maltho	139
Notropis (231)	23
longurio, Carcharhinus (43)	8
Carcharias	8
longus, Ophisurus	53
Lophiidæ (Family CXLIX)	138
Lophius piscatorius (1630)	138
radiatus	139
vespertilio	138
lophius, Amiurus	15
Lophobranchii	61
Lopholatilus chamæleonticeps (1214)	104
Lota lota maculosa (1542)	129
lota, Lota (1542)	129
Lotella maxillaris (1552)	130
schlegeli	130
Lotinæ	128
louisianæ, Siphostoma (691)	62
Lucania goodei (584)	49
parva (583)	49
venusta (582)	49
lucasana, Julis	90
lucasanum, Thalassoma (1166)	99
lucens, Ceratichthys	24, 48
luciæ, Zygonectes (581)	49
lucidus, Luxilus	26
Stolephorus (470)	38
lucioceps, Synodus (480)	30
lucius, Esox (600)	51
Ptychochilus (357)	30
ludibundus, Notropis (255)	25
lugubrosus, Esox	50
lumbricus, Myrophis (629)	54
Lumpenus anguillaris (1490)	123
lumpenus (1491)	123
medius (1489)	123
lumpenus, Lumpenus (1491)	123

	Page.
lumpus, Cyclopterus (1410)	116
lunulatus, Galeus (25)	6
Mustelus	6
lupus, Amiurus (130)	15
Anarrhichas (1498)	123
lusca, Cyclothone (537)	46
lutea, Perca (947)	81
luteovinctum, Etheostoma (930)	80
lutipinnis, Notropis (279)	26
Opisthopterus (454)	37
Pristigaster	37
Lutjanus	86
analis (1014)	87
aratus (1016)	87
argentiventris (1006)	87
blackfordi	87
caxis (1007)	87
colorado (1015)	87
griseus (1009)	87
guttatus (1011)	87
inermis (1017)	87
jocu (1008)	87
novemfasciatus (1010)	87
prieto	87
stearnsi	87
synagris (1012)	87
vivanus (1013)	87
Lutodeira	35
lutrensis, Leuciscus	24
Notropis (240)	24
luxatus, Chasmistes (174)	18
Luxilinus occidentalis (416)	33
Luxilus	26, 33
chickasavensis	25
lucidus	26
selene	24
Lycenchelys paxilloides (1508)	124
paxillus (1507)	124
verrilli (1509)	124
Lycocara parrii (1520)	125
Lycodalepis mucosus (1516)	125
polaris (1518)	125
turneri (1517)	125
Lycodes	125
coccineus (1515)	124
esmarki (1511)	124
muræna	124
nebulosus (1514)	124
paxillus	124
reticulatus (1512)	124
seminudus (1513)	124
vahli (1510)	124
Lycodidæ (Family CXXXIX)	124, 125, 126
Lycodonus mirabilis (1506)	124
Lycodopsis pacificus (1504)	124
pancidens (1505)	124
lynceum, Etheostoma (917)	80
Lyomeri	57
Lyopsetta	135
lyricus, Gobius (1225)	105
Lythrulon	89
Lythrurus	26
lythrurus var. (276 b)	26
Notropis	26
macarellus, Decapterus (778)	70

CATALOGUE OF THE FISHES OF NORTH AMERICA.

	Page.
macellus, Prionistius (1355)	112
mackayi, Siphostoma (693)	62
mackenziei, Stenodus (517)	43
maclura, Pteroplatea (83)	11
macracanthus, Pomadasys (1033)	89
macrocephalus, Hadropterus (900)	79
macrochilus, Catostomus (163)	17
macrochir, Halosaurus (4-8)	41
macrochirus, Lepomis (857)	77
Macrodonophis mordax	53
macrolepidotum, Moxostoma (185)	19
macrolepidotus, Engraulis	37
Notropis	26
Pogonichthys (350)	30
Stolephorus (458)	37
macrophthalma, Clupea	36
macrophthalmus, Anthias	86
Priacanthus	86
macropoma, Centropristis	82
macrops, Citharichthys (1586)	133
Hippoglossina (1604)	135
macropterus, Centrarchus (841)	76
Macrorhamphosidæ (Family LXX)	62
Macrorhamphosus scolopax (701)	62
macrostomus, Notropis (257)	25
Macruridæ (Family CXLVI)	131
Macrurus	129
acrolepis (1569)	131
asper (1572)	131
bairdii (1571)	131
berglax (1568)	131
carminatus (1570)	131
fabricii	131
rupestris	131
macrurus, Ophichthys (623)	53
macularius, Cyprinodon (552)	47
maculatum var. (885 d)	78
Aulostoma (705)	63
Boleosoma	78
Cynoscion (1120)	95
Etheostoma (921)	80
maculatus, Alvordius	79
Apogon (1074)	92
Bothus (1576)	132
Cryptacanthodes (1497)	123
Dormitator (1223)	105
Galeocerdo (31)	7
Galeus	7
Hadropterus	79
Heptranchias (13)	4
Leptoclinus (1488)	123
Notropis (225)	22
Rhypticus	86
Scomberomorus (767)	68
Upeneus (1079)	93
maculicauda, Hæmulon (1040)	89
maculipinna, Julis	90
Platyglossus (1161)	99
maculocinctus, Chætodon (1199)	102
maculofasciatus, Serranus (967)	83
maculosa var. (1542)	129
maculosus, Oligocottus (1363)	113
mæandricus, Gobiesox (1411)	116
majalis, Fundulus (557)	48
major, Liparis (1397)	115

	Page.
Makaira nigricans	67
Malacanthidæ (Family CXX)	104
Malacanthus	104
Malacosteus niger (492)	42
maliger, Sebastichthys (1287)	108
Mallotus villosus (495)	42
malma, Salvelinus (529)	44
Malthe cubifrons	8, 139
elater (1652)	139
longirostris	139
radiata	8
vespertilio (1651)	139
radiata (1651 b)	139
Malthidæ (Family CLII)	139
manatinus, Barathrodemus (1534)	127
Mancalias uranoscopus (1647)	138
maniton, Percina	79
Manta birostris (97)	12
marconis, Hybopsis (339)	29
margaritatus, Porichthys (1420)	116
margaritus, Phoxinus (378)	31
marginata, Uranidea (1325)	111
marginatum, Ophidion (1525)	126
marginatus, Dinematichthys (1532)	127
Lepomis (866)	77
marinus, Ælurichthys (141)	16
Petromyzon (11)	3, 4
Sebastes (1262)	107
Tylosurus (660)	59
marmorata, Pteroplatea (84)	11
marmoratus var. (125 c)	15
Amiurus	15
Cremnobates (1465)	121
Scorpænichthys (1361)	112
Marsipobranchii	3
martinicus, Upeneus (1080)	93
Mascalongus	51
Mastacembelidæ	58
matejuelo, Amphiprion	75
Holocentrum	75
matutinus, Notropis (301)	27
matzubaræ, Sebastichthys (1275)	107, 108
Maurolicus borealis (487)	40
maxillaris, Lotella (1552)	130
Muræenoides (1474)	122
maxillingua, Exoglossum (220)	22
maxillosus, Gnathypops (1433)	118
maximus, Cetorhinus (53)	9
Labrus	97
Lachnolæmus (1152)	97
Maynea	125
mazatlana, Seriola (806)	72
Solea	137
mazatlanus, Achirus (1633)	137
Meda argentissima (424)	33
fulgida (423)	33
mediocris, Clupea (443)	36
medirostris, Acipenser (103)	13
mediterraneus, Hoplostethus (833)	75
medius, Centropomus	82
Lumpenus (1489)	123
Stromateus (817)	73
meeki, Hybognathus (214)	21
megacephalus, Chitonotus (1310)	110
megalepis, Doratonotus	99

	Page.		Page.
Megalops atlanticus (434)	34	Microdesmus retropinnis	126
notata	36	microdon, Pseudotriacis (23)	6
oglina	36	Pseudotrakis	6
megalops, Alburnellus	26	Microgadus proximus (1559)	130
Cyprinus	26	tomcod (1560)	130
Notropis (273)	26, 27	Microlepidotus inornatus	88
Trycherodon	33	microlepidotus, Orthodon (260)	20
megalotis, Lepomis (864)	77	Prionurus	103
Melamphaës	74	microlepis, Mycteroperca (979)	84
melanogaster, Pleuronectes	134	Micrometrus aggregatus (1137)	96
Melanogrammus æglefinus (1555)	130	micronemus, Peristedion	114
melanopoma, Polynemus	66	Microperca fonticola	81
melanops, Dionda (206)	21	microphthalmus, Dormitator	105
Haplochilus	50	Micropogon cetoneus (1100)	94
Minytrema (177)	19	undulatus (1099)	94
Sebastichthys (1265)	107	micropogon, Ceratichthys	28
melanostictus, Psettichthys (1609)	135	microps, Caulolatilus (1216)	104
Melanostigma gelantinosum (1521)	125	Cottunculus (1303)	110
melanura, Nettastoma	54	Micropterus dolomiei (877)	77
melanurum, Nettastoma	55	salmoides (876)	77
melanurus, Exocœtus	61	micropteryx, Notropis (311)	27
tuckas, Amiurus (124)	14	micropus, Etrumeus	35
melastomus	6	microstigmius, Myrophis	54
Meletta libertatis	37	Microstoma groenlandicum (494)	42
Melletes papilio (1360)	112	microstomus, Citharichthys (1589)	133
Menidia audens (732)	65	Minnilus	23
beryllina (733)	65	milneri, Nocomis	29
bosci	65	Pagellus	91
laciniata (729)	65	milnerianus, Phoxinus (404)	31
menidia (734)	65	miniatum, Peristedium (1382)	114
notata (731)	65	miniatus, Lepomis (862)	77
peninsulæ (735)	65	Sebastichthys (1274)	108
vagrans (730)	65	Miniellus	26
menidia, Menidia (734)	65	minima, Abeona (1133)	96
menona, Fundulus	49	Minnilus	22
Menticirrus	33	dinemus	27
alburnus (1109)	94	diplæmius	26
elongatus (1106)	94	microstomus	23
littoralis (1105)	94	nigripinnis	26
nasus (1111)	94	rubripinnis	27
nebulosus	94	minor, Anarrhichas (1499)	123
panamensis (1110)	94	minuta, Uranidea (1322)	111
saxatilis (1108)	94	Minytrema melanops (177)	19
undulatus (1107)	94	mirabilis, Clupea (439)	35
meridionalis var. (1320 f)	111	Gillichthys (1237)	106
merki, Coregonus (511)	43	Lycodonus (1506)	124
Merlucius bilinearis (1565)	131	Phenacobius (316)	27
merlucius (1566)	131	missuriensis, Chola	23
productus (1567)	131	mitchilli, Stolephorus (466)	38
merlucius, Merlucius (1566)	131	mitis, Balistes	140
mesoum var. (885 f)	78	miurus, Noturus (116)	14
mesogaster, Exocœtus	60, 61	Ophichthys (619)	53
Parexocœtus (671)	60	modestus, Anisotremus	89
Mesogonistius chætodon (852)	76	Phoxinus (401)	31
Mesoprion argentiventris	87	Pseudojulis (1165)	99
campechanus	87	Mola mola (1685)	141
vivanus	87	mola, Mola (1685)	141
metallica, Agosia (323)	28	Molacanthus nummularis	141
metallicus, Notropis (303)	27	Mollienesia latipinna (591)	50
mexicanum, Dorosoma (456)	37	lineolata	50
mirchus, Stolephorus (468)	38	mollis var. (1635 b)	137
Micristodus punctatus (54)	10	Molva molva (1553)	130
microcephalus, Gasterosteus (710)	63	molva, Molva (1553)	130
Somulosus (17)	5	Monacanthus ciliatus (1663)	140
Microdesmus dipus (1522)	125, 126	davidsoni	140

CATALOGUE OF THE FISHES OF NORTH AMERICA. [167]

	Page.
Monacanthus hispidus (1664)	140
occidentalis	140
pullus (1666)	140
spilonotus (1665)	140
monachus, Hybopsis (334)	20
monæ, Stephanoberyx (828)	74
Moniana	24
aurata	25
complanata	24
couchi	24
deliciosa	23
frigida	24
gibbosa	24
gracilis	24
lætabilis	24
leonina	24
nitida	23
proserpina	25
pulchella	24
rutila	24
Monochir pilosus	137
reticulatus	137
Monolene sessilicauda (1630)	136
Monomitra liparina (1394)	115
monopterygius, Aspidophoroides (1369)	113
Pleurogrammus (1251)	106
montagnei, Liparis (1403)	115
montanus, Hybopsis (338)	20
Phoxinus (372)	30
Mora	129
mordax, Crotalopsis	53
Macrodonophis	53
Ophichthys	53
Osmerus (498)	42
Sidera (607)	51
moringa, Sidera (611)	52
morio, Epinephelus (983)	84
Morone	82
morrisi, Leptocephalus	55
Motella septentrionalis	128
Moxostoma album (182)	19
anisurum (190)	19, 20
aureolum (186)	19, 20
bucco	19
cervinum (192)	20
claviformis	19
congestum (188)	19
conus (189)	20
coregonus (181)	19
crassilabre (187)	19
kennerlyi	19
macrolepidotum (185)	19
duquesnei (185 b)	19
papillosum (178)	19
pidiense (180)	19
pœcilurum (191)	20
thalassinum (183)	19
valenciennesi (184)	19
velatum (179)	19
mucosa, Liparis (1406)	115
mucosus, Lycodalepis (1516)	125
Xiphister (1481)	122
mucronatus, Neoconger (633)	54
Mugil albula	64
brasiliensis	64

	Page.
Mugil cephalus (715)	64
chanos	35
curema (717)	64
gaimardianus (716)	64
liza	64
salmoneus	35
trichodon (718)	64
Mugilidæ (Family LXXV)	64
mülleri, Myctophum (485)	40
Salmo	40
Scopelus	40
Mullidæ (Family CVIII)	92
Mulloides flavovittatus	93
Mullus barbatus auratus (1078)	92
multifasciata, Adinia (556)	48
multifasciatus, Anthias (971)	83
Pronotogrammus	83
multifilis, Hyplenrochilus (1448)	119
multiguttatum, Plectropoma	84
multiguttatus, Alpheestes (991)	84
mundiceps, Xyrichthys (1169)	100
mundicorpus, Iniistius	100
Novacula	100
Xyrichthys (1170)	100
mundus, Dactyloscopus (1424)	117
Muræna afra	52
dovii	51
funebris	52
infernalis	52
lineopinnis	52
pinta (605)	51
pintita	51
retifera (604)	51
muræna, Lycodes	124
Muræenesox coniceps (635)	55
Muræenidæ (Family LVIII)	51
Muræenoblenna nectura (603)	51
olivacea	51
Muræenoides dolichogaster (1475)	122
fasciatus (1472)	122
gunelliformis	122
gunnellus (1471)	122
maxillaris (1474)	122
ornatus (1473)	122
murinus, Lepomis (860)	77
Mustelus lunulatus	6
mustelus, Galeus	6
Squalus	6
Mycteroperca bonaci (980)	84
xanthosticta (980 b)	84
falcata phenax (978)	84
microlepis (979)	84
rosacea (977)	84
venenosa (981)	84
Myctophum boops (466)	40
crenulare (484)	39
mülleri (485)	40
mydrus, Cyprinodon (553)	47
Myliobatidæ (Family XXIII)	12
Myliobatis californicus (95)	12
fremnvillei (94)	12
Mylochilus caurinus (352)	30
Myloleucus	32
parovanus	32
thalassinus	32

	Page.
Mylopharodon conocephalus (353)	30
myops, Synodus (482)	39
Myrichthys tigrinus (628)	54
Myriolepis zonifer (1200)	107
Myriopristis occidentalis	76
pœcilopus	76
Myripristis jacobus	75
occidentalis (836)	75
pœcilopus (837)	75
Myrophis egmontis (662)	54
longicollis	54
lumbricus (629)	54
microstigmius	54
punctatus (630)	54
vafer (631)	54
mystacalis, Lepomis (858)	77
mystacinus, Gnathypops (1432)	118
mystinus, Sebastichthys (1267)	107
Myxine glutinosa (2)	3
Myxinidæ (Family II)	3
Myxodagnus opercularis (1427)	117
Myxodes	117
Myxus harengus	64
nacrea, Gila (364)	30
namaycush, Salvelinus (526)	44
Nannostomus	80
nanomyzon, Catostomus	17
Nanostoma elegans	80
Narcine brasiliensis (78)	11
corallina (78 b)	11
umbrosa (79)	11
naresi, Salvelinus	44
narinari, Stoasodon (92)	12
nasus, Menticirrus (1111)	94
Umbrina	94
nasutus, Agonostomus (722)	64
Trachynotus	71
natalis, Amiurus (127)	15
Naucrates	60
ductor (803)	71
Echeneis (750)	66
Nautichthys oculofasciatus (1367)	113
navaga, Gadus	130
Pleurogadus (1558)	130
nebularis, Platophrys (1378)	132
nebulifer, Catostomus (158)	17
Serranus (968)	83
nebulosa, Aphoristia (1638)	137
nebulosus, Amiurus (125)	14, 15
Lycodes (1514)	124
Menticirrus	94
Sebastichthys (1269)	108
nectura, Gymnomuræna	51
Muroenoblenna (603)	51
nelsoni, Coregonus (507)	43
Nematistius	69
pectoralis (811)	72
Nematognathi	14
Nemichthyidæ (Family LXIII)	56
Nemichthys avocetta (643)	56
scolopaceus (642)	56
Neoclinus blanchardi (1458)	120
satiricus (1457)	120
Neoconger mucronatus (633)	54
neogæus, Phoxinus (402)	31

	Page.
Neoliparis	115
nephelus, Tetrodon (1673)	141
nerka, Oncorhynchus (522)	44
Nestis	64
Nettastoma melanura	5
melanurum	55
procerum (634)	54, 55
Netuma	15
novisense, Etheostoma	79
newberryi, Lepidogobius (1239)	106
nianguæ, Hadropterus (911)	79
nicholsi, Gobius (1233)	105
niger, Astronesthes (493)	42
Chiasmodon (1437)	119
Cottus (1345)	111
Malacosteus (492)	42
Petromyzon	4
Phoxinus (392)	31
nigra, Etheostoma	78
nigrescens, Centropomus (951)	82
Coryphæna	82
Phoxinus (400)	31
Tigoma	31
nigricans, Amiurus (132)	15
Catostomus	18
Girella (1068)	91
Hypentelium (171)	18
Hypoplectrus (969)	83
Makaira	67
nigrilabris, Gronias (121)	14
nigripinnis, Coregonus (514)	43
Minnilus	26
Rhypticus (999)	86
nigrirostris, Chætodon (1203)	102
Sarothrodus	102
nigritus, Epinephelus (982)	84
nigrocinctus, Sebastichthys (1291)	108
nigrofasciatus, Fundulus	40
Hadropterus (907)	79
nigrotæniatus, Hybognathus	21
nitida, Moniana	23
nitidus, Notropis (232)	23
Pomadasys (1029)	88
Salvelinus	44
niveatus, Epinephelus (986)	84
niveiventris, Amiurus (131)	15
niveus, Notropis (265)	25
nivipes, Emblemaria (1456)	120
nobile, Cynoscion (1112)	95
nobilior, Esox (601)	51
nobilis, Couodon (1020)	88
Gambusia (589)	50
Nocomis	28
hyostomus	29
milneri	29
nocomis, Notropis (237)	24
nocturnus, Noturus (113)	14
Nomeidæ (Family LXXXVII)	72
Nomeus gronovii (815)	72
normalis, Bassozetus (1536)	128
notabilis, Argyreus	28
Notacanthidæ (Family LXVII)	58
Notacanthus analis (653)	58
chemnitzi (651)	58
phasganorus (652)	58

[169] CATALOGUE OF THE FISHES OF NORTH AMERICA.

	Page.
Notarius	15
notata, Cyprinella	25
Megalops	36
Menidia (731)	65
notatum, Pristipoma	80
notatus, Lepomis (874)	77
Notropis (258)	25
Pimephales (219)	22
Porichthys	116
Tylosurus (658)	59
Zygonectes (576)	40
Notemigonus chrysoleucus (418)	33
bosci (418 b)	33
gardonens (417)	33
Nothonotus	80
nothum, Cynoscion (1115)	95
Notidanidæ (Family V)	4
Notogrammus rothrocki (1487)	123
Notorhynchus	4
Notosema dilecta	134
notospilotus, Artedius (1306)	110
notospilus, Pseudojulis (1164)	99
Notropis	21
alabamæ	27
altipinnis (291)	26
amabilis (292)	26
amarus	24, 28
analostanus	25
anogenus (227)	23
ardens (296)	26
atripes (296 c)	26
cyanocephalus (296 d)	26
lythrurus (296 b)	26
ariommus (286)	26
atherinoides (308)	27
bellus (300)	27
bifrenatus (224)	22
bivittatus (295)	26
blennius (244)	23, 24
boops (243)	24
bubalinus (253)	25
callisema (252)	25
callistius (266)	25
camurus (263)	25
cercostigma (260)	25
stigmaturus (260 b)	25
chalybæus (282)	26
chiliticus (281)	26
chloristius (269)	25
chlorocephalus (280)	26
chlorus (239)	24
chrosomus (283)	26
coccogenis (274)	24, 26
cœruleus (268)	25
comalis (240)	24
deliciosus (233)	23
longiceps (233 c)	23
stramineus (233 b)	23
voluncellus (233 d)	23
dilectus (309)	27
domninus (312)	27
eurystomus (264)	25
formosus (251)	25
fretensis (230)	23
galacturus (262)	25

	Page.
Notropis garmani (256)	25
gilberti (235)	23
heterodon (226)	22
hudsonius (246)	24
amarus (246 b)	24
hypselopterus (272)	25
illecebrosus (229)	23, 24
jejunus (288)	26
lacertosus (285)	26
leoninus (248)	24
lepidus (254)	25
lenciodus (289)	26
lirus (302)	27
longirostris (231)	23
ludibundus (255)	25
lutipinnis (279)	26
lutrensis (249)	24
lythrurus	26
macrolepidotus	26
macrostomus (257)	25
maculatus (225)	22
matutinus (301)	27
megalops (273)	26, 27
cyaneus (273 c)	26
frontalis (273 b)	26
metallicus (303)	27
micropteryx (311)	27
nitidus (232)	23
niveus (265)	25
nocomis (237)	24
notatus (258)	25
ornatus (247)	24
phenacobius (238)	24
photogenis (305)	27
piptolepis (241)	24
procne (234)	23
proserpina (250)	25
punctulatus (298)	27
pyrrhomelas (271)	25
roseipinnis (299)	27
roseus (277)	26
rubricroceus (278)	26
rubrifrons (310)	27
scabriceps (287)	26
scepticus (334)	27
scylla (236)	24
simus (245)	24
socius (293)	26
spectrunculus (229)	23
spilurus (290)	26
stilbius (307)	27
stramineus	23
swaini (294)	26
telescopus (306)	27
timpanogensis (313)	27
topeka (242)	24
trichroistius (267)	25
umbratilis (297)	27
venustus (259)	25
whipplei (261)	25
xænocephalus (284)	26
xænurus (270)	25
zonatus (275)	26
zonistius (276)	26
Noturus elassochir	14

	Page.
Noturus eleutherus	14
exilis (117)	14
flavus (119)	14
funebris (114)	14
gyrinus (111)	14
insignis (118)	14
latifrons (115)	14
leptacanthus (112)	14
miurus (116)	14
nocturnus (113)	14
Novacula mundicorpus	100
Novaculichthys callosoma	100
novemfasciatus, Lutjanus (1010)	87
novemradiata, Agosia (324)	28
nox, Cremnobates (1460)	121, 122
nubila, Agosia (326)	28
Apocope	28
Cliola	21
Dionda (212)	21
nubilus, Leptoblennius (1407)	123
nuchalis, Ælurichthys	16
Hybognathus (216)	21
nuchipinnis, Labrosomus (1459)	120
nummularis, Molacanthus	141
obesa, Algansea (400)	32
obesus, Amiurus	14
Enneacanthus (849)	76
Phoxinus (386)	31
oblonga, Platessa	134
oblongus var. (176 b)	19
Paralichthys (1000)	134
Pleuronectes	134
Pseudorhombus	134
obscuratus, Pomacentrus (1184)	101
obscurus, Carcharhinus (33)	7
obtusirostris, Exocœtus	60
occidentalis, Catostomus (164)	17, 18
Heterandria (594)	50
Luxilinus (416)	33
Luxilus	33
Monacanthus	140
Myriopristis	76
Myripristis (836)	75
Torpedo (76)	11
occipitalis, Scorpæna (1298)	109
oceanicus, Gobionellus (1235)	104, 106
ocellaris, Fundulus (567)	40
Platessa	134
Pseudorhombus	134
ocellata, Raia (64)	11
Sciæna (1091)	93
Sidera (609)	51
ocellatus var.	138
Anarrhichthys (1502)	123
Antennarius (1642)	138
Chætodon (1200)	102
Citharichthys (1579)	133
Ophichthys (621)	53
Rhombus	132
Xenopsis (827)	74
octodecimspinosus, Cottus (1333)	111
octofilis, Polynemus	66
octonemus, Polynemus (746)	66
oculofasciatus, Nautichthys (1367)	113
Ocyurus chrysurus (1018)	87

	Page.
Odontaspididæ (Family XIII)	9
Odontaspis	9
taurus	7
Odontopyxis trispinosus (1378)	114
œrstedi, Solene (792)	71
ogac, Gadus (1557)	130
oglina, Megalops	36
oglinum, Opisthonema (451)	36
olfersi, Argyropelecus (534)	45
Pleurothyris	45
olidus, Hypomesus (501)	42
Oligocottus analis (1362)	113
globiceps (1364)	113
maculosus (1363)	113
Oligoplites	70
altus (812)	72
saurus (813)	72
olisthostoma, Gerres (1124)	95, 96
olivacea, Algansea (412)	32
Murænoblenna	51
olivaceus, Leucus	32
olivaris, Leptops (120)	14
olmstedi, Boleosoma (885)	78
olriki, Aspidophoroides (1371)	113
ommata, Heterandria (595)	50
omostigma, Genypterus	126
Otophidium (1529)	126
Oncorhynchus gorbuscha (518)	43
keta (519)	44
kisutch (521)	44
nerka (522)	44
tchawytcha (520)	44
Oneirodes eschrichti (1648)	139
Oninæ	128
onitis, Hintula (1151)	97
Onos ensis (1539)	128
reinhardti (1538)	128
rufus (1540)	128
septentrionalis (1541)	128
ontariensis var. (516 b)	43
Thymallus	43
opercularis, Myxodagnus (1427)	117
Polynemus (745)	66
Stolephorus (459)	37
Ophichthys	52
chrysops (624)	53
guttifer (622)	53
intertinctus (627)	53
macrurus (623)	53
miurus (619)	53
mordax	53
ocellatus (621)	53
punctifer	53
schneideri (626)	53
triserialis (620)	53
xysturus	53
zophochir (625)	53
Ophidiidæ (Family CXLIII)	126
Ophidion beani (1527)	126
holbrooki (1526)	126
marginatum (1525)	126
Ophidium gracilki	126
josephi	126
parrii	125
Ophioblennius webbi (1438)	119

CATALOGUE OF THE FISHES OF NORTH AMERICA.

	Page.
Ophiodon elongatus (1257)	107
Ophisuraphis	52
Ophisurus acuminatus (617)	53
interiuctus	54
longus	53
xysturus (618)	53
ophryas, Paralichthys	134
Prionotus (1387)	115
Opisthartbri	4
Opisthognthidæ (Family CXXXIV)	118
Opisthognathus	104
lonchura (1435)	118
punctata (1436)	118
rhomaleus	118
scaphiura (1434)	118
Opisthomi	58
Opisthonema libertate (452)	37
oglinum (451)	36
Opisthopterus lutipinnis (454)	37
Opsanocodus emiliæ (415)	33
oquassa, Salvelinus (527)	44
Orcynus alalonga (773)	69
thynnus (774)	69
ordinatus, Hexagrammus (1252)	106
oreas, Chrosomus (203)	20
oregonensis, Ptychocheilus (354)	30
ornata var. (67)	11
ornatum, Campostoma (195)	20
ornatus, Cochlognathus (221)	22
Murænoides (1473)	122
Notropis (247)	24
Ornichthys	114
orqueta, Chloroscombrus (795)	71
Orthagoriscidæ (Family CLVII)	141
Orthagoriscus	141
Orthodon microlepidotus (200)	20
Orthopristis brevipinnis (1023)	88
cantharinus (1024)	88
chalceus (1025)	88
chrysopterus (1026)	88
inornatus (1022)	88
Orthostœchus	89
oscula, Agosia (327)	28
osculus, Argyreus	28
osmerinus, Hybognathus	21
Osmerus attenuatus	42
dentex (499)	42
mordax (498)	42
thaleichthys (497)	42
osseus, Lepidosteus (107)	13
osteochir, Rhombochirus (735)	66
Ostraciidæ (Family CLIII)	139
Ostracion quadricornis	139
tricorne (1657)	139
trigonum (1656)	139
triquetrum (1655)	139
othonopterum, Cynoscion (1116)	95
Otolithus reticulatus	95
Otophidium omostigma (1529)	126
taylori (1528)	126
otrynter, Caranx	70
ouachitæ, Hadropterus (903)	79
ovalis, Citharichthys (1581)	133
Hemirhombus	133
Sebastichthys (1269)	107

	Page.
oxygeneios, Epinephelus	83
Oxygeneum pulverulentum (198)	20
Oxyjulis	90
Oxylebius pictus (1259)	107
oxyrhynchus var. (101)	13
Tetrodon	141
ozarcanum var. (885 e)	78
pacificus, Cynicoglossus (1628)	136
Lycodopsis (1504)	124
Thaleichthys (496)	42
pætulus, Citharichthys (1580)	133
Hemirhombus	133
Pagellus milneri	91
penna	91
pagrus, Sparus (1054)	90
pallidus, Lepomis (869)	77
Platygobio	29
Pomotis	77
palmipes, Prionotus (1385)	114
palustris, Pœcilichthys	81
panamensis, Æluriichthys (142)	16
Caranx	70
Citharichthys (1582)	133
Menticirrus (1110)	94
Pomadasys (1031)	89
Umbrina	94
pandionis, Apogon (1077)	92
pandora, Phoxinus (377)	31
Pantosteus bardus	17
delphinus	17
generosus (152)	17
guzmaniensis (153)	17
platyrhynchus	17
plebeius (151)	17
virescens	17
papilio, Melletes (1360)	112
papillifer, Chologaster (543)	47
papillosum, Moxostoma (178)	19
paradoxus, Psychrolutes (1302)	109
Paralabrax	83
Paralepididæ (Family XLIII)	38
Paralichthys	135
adspersus (1594)	133
albigutta (1598)	134
californicus (1595)	133
dentatus (1596)	134
lethostigma (1597)	134
oblongus (1600)	134
ophryas	134
squamilentus (1599)	134
Paranthias furcifer (973)	83
parasiticus, Simenchelys (639)	56
pardus var. (1419 b)	116
Pareques	94
Parexocœtus mesogaster (671)	60
parietalis, Coliscus	22
parmatus, Setarches (1299)	109
parnifera, Raia (70)	11
Parophrys ischyrus	136
vetulus (1614)	135
parovana, Algansea (400)	32
parovanus, Mylolcucus	32
parrii, Lycocara (1520)	125
Ophidium	125
paru, Stromateus (816)	72

	Page.
parva, Lucania (583)	40
parvipinne, Cynoscion (1117)	95
Etheostoma (931)	80
parvipinnis, Fundulus (559)	48
patronus var. (153 b)	37
patruelis, Gambusia (585)	50
paucidens, Lycodopsis (1505)	124
paucispinis, Sebastodes (1263)	107
pavo, Xyrichthys	100
pavonaceus, Heros (1183)	101
paxilloides, Lycenchelys (1508)	124
paxillus, Lycenchelys (1507)	124
Lycodes	124
pectinatus, Pristis (56)	10
pectoralis, Bodianus (1155)	97
pectoralis, Dactyloscopus (1425)	117
Dallia (602)	51
Harpe	97
Nematistius (811)	72
Pediculati	138
pedimacula, Centropomus (952)	82
pelamys, Euthynnus (776)	69
pelecanoides, Eurypharynx	58
pellucida, Ammocrypta (880)	77
pellucidus, Delothyris (1629)	136
Thyris	136
peltatum, Etheostoma	79
peltatus, Hadropterus (904)	79
peninsulæ, Menidia (735)	65
penna, Calamus (1060)	91
Pagellus	91
pennatula, Calamus	90
pensacolæ, Clupea (449)	36
pentacanthum, Holocentrum	75
Perca ascensionis	75
chrysoptera	88
furva	82
gibbosa	90
lutea (947)	81
philadelphica	82
saxatilis	82
sectatrix	92
septentrionalis	82
trifurca	82
unimaculata	91
variabilis	108
venenosa	84
Percesoces	64
Percidæ (Family XCIX)	77
perciformis, Leirus (820)	73
Percina caprodes (899)	79
zebra (899 b)	79
maniton	79
percobromus, Alburnellus	27
Percomorphi	66
Percopsidæ (Family L)	44
Percopsis guttatus (532)	44
perfasciatus, Engraulis	38
Stolephorus (463)	37, 38
Peristedion micronemus	114
Peristedium imberbe (1383)	114
miniatum (1382)	114
perrico, Scarus (1181)	101
perrottetii, Pristus (57)	10
personatus var. (747 b)	66

	Page.
perthecatus, Stolephorus (461)	37
peruvianus, Gerres (1125)	95
Petrometopon	85
Petromyzon bairdii	4
bdellium (8)	4
castaneus (10)	4
hirudo (9)	4
marinus (11)	3, 4
dorsatus (11 b)	4
niger	4
plumbeus	4
Petromyzontidæ (Family IV)	3
petrosus, Serranus	84
petus, Acanthocybium	68
Cybium	68
Phanerodon	97
Pharyngognathi	66
phasganorus, Notacanthus (652)	58
Phenacobius catastomus (317)	27
mirabilis (316)	27
teretulus (315)	27
uranops (318)	27
phenacobius, Notropis (238)	24
phenax var	84
Lepomis (855)	77
philadelphica, Perca	82
philadelphicus, Serranus (960)	82
philippi, Cestracion	5
Philypnus lateralis	104
phlebotomus, Acanthurus	103
phlegethontis, Phoxinus (405)	31
phlox, Ulocentra (889)	78
phœbe, Centropristis	83
Haliperca	83
Serranus (964)	82
Pholidichthys anguilliformis (1495)	125
leucotœnia (123)	123
Pholis	110
Photogenis piptolepis	24
stigmaturus	25
photogenis, Notropis (305)	27
Phoxinus	27
aliciæ (390)	31
ardesiacus (376)	31
atrarius (395)	31
bicolor (385)	31
cœruleus (308)	31
conformis (384)	31
conspersus (393)	31
cooperi (390)	31
copei (391)	31
crassicauda (394)	31
crassus (397)	31
cruorens (375)	13
egregius (381)	31
elongatus (368)	30
estor (368)	30
flammeus (405)	31
funduloides (369)	30
galtiæ (374)	31
gracilis (383)	31
gula (370)	31
humboldti (373)	30
hydrophlox (370)	30
intermedius (389)	31

[173] CATALOGUE OF THE FISHES OF NORTH AMERICA.

	Page.
Phoxinus lineatus (382)	31
margaritus (378)	31
milnerianus (404)	31
modestus (401)	31
montanus (372)	30
neogæus (402)	31
niger (392)	31
nigrescens (400)	31
obesus (386)	31
pandora (377)	31
phlegethontis (405)	31
pulchellus (388)	31
pulcher (380)	31
purpureus (387)	31
squamatus (396)	31
tænia (371)	30
vandoisulus (367)	30
phoxocephalus, Hadropterus (901)	70
Phthcirichthys lineatus (751)	66
Phycis chesteri (1548)	129
chuss (1546)	129
earlli (1545)	129
floridanus (1544)	129
regius (1543)	129
tenuis (1547)	129
yarrelli	129
Physiculus dalwigkii	130
fulvus (1551)	130
physignathus, Conesius (345)	29
picarti, Hemirhamphus	60
Picorellus	50
picturatus, Trachurus (770)	70
pictus, Chaunax (1645)	138
Oxylebius (1259)	107
picuda, Sphyræna (741)	65
pidiense, Moxostoma (180)	19
Pikeoma zebra	70
pilosa, Solea	137
pilosus, Monochir	137
Trichodiodon (1677)	141
Pimelepteridæ	92
Pimelepterus analogus	92
bosci	92
Pimelodus catulus	14
Pimelometopon	98
Pimephales notatus (219)	22
promelas (218)	22
confertus (218b)	22
pingeli, Triglops (1334)	112
pinnatus, Synaphobranchus (640)	56
pinniger var. (851 b)	76
Sebastichthys (1273)	107, 108
pinnimaculatus, Ælurichthys (143)	16
pinnulatus, Elagatis (810)	72
pinta, Muræna (605)	51
pintia, Muræna	51
piptolepis, Notropis (241)	24
Photogenis	24
piscatorius, Lophius (1630)	138
Pisces	4
pisonis, Eleotris (1220)	105
pistilliger, Gymnacanthus (1348)	112
pituitosus, Rhypticus	66
placita var	21
Hybognathus	21

	Page.
Placopharynx cariuatus (193)	20
plagiusa, Aphoristia (1637)	137
Plagopterus	33
Plagyodus æsculapius (473)	38
borealis (474)	38
ferox (472)	38
Platessa oblonga	134
ocellaris	134
platessoides, Hippoglossoides (1606)	135
Platichthys	136
Platophrys	133, 136
leopardinus (1577)	132
nebularis (1578)	132
platycephalus, Amiurus (123)	14
Cottus (1343)	111
Platyglossus bivittatus (1159)	98
caudalis (1160)	98
cyanostigma	98
dispilus (1163)	99
florealis	98
grandisquamis	98
humeralis	98
maculipinna (1161)	99
radiatus (1158)	98
semicinctus (1162)	99
Platygobio gracilis (346)	29
pallidus	29
platyodon, Carcharhinus (36)	7
platypogon, Arius	16
Galeichthys (139)	16
Platyrhinoidis	10
platyrhynchus, Pantosteus	17
Scaphirhynchops (106)	13
Platysomatichthys	133
platystomus, Lepidosteus (108)	13
plebeius, Pantosteus (151)	17
Plectognathi	139
plectrodon, Porichthys	116
Plectromus crassiceps (831)	74
suborbitalis (830)	74
Plectropoma multiguttatum	84
pleei, Hemirhamphus (66e)	60
Pleuracromylon	7
Pleurogadus navaga (1558)	130
Pleurogrammus monopterygius (1251)	106
Pleurolepis asprellus	78
Pleuronectes achirus	137
americanus (1625)	136
dentatus	134
glaber (1623)	136
glacialis (1624)	136
lineatus	137
melanogaster	134
oblongus	134
quadrituberculatus (1622)	136
stellatus (1621)	136
Pleuronectidæ (Family CXLVII)	132
Pleuronichthys cœnosus (1612)	135
decurrens (1610)	135
verticalis (1611)	135
pleurophthalmus, Antennarius	138
Pleurothyris olfersi	45
plumbea, Chimæra	12
plumbeolus, Hybopsis	26
plumbeum, Zophendum (205)	20

	Page.
plumbeus, Cottus (344)	29
Gobio	29
Petromyzon	4
plumieri, Gerres (1122)	95
Hæmulon (1046)	90
Polydactylus	66
Scorpæna (1295)	8, 109
plutonia, Raia (68)	11
Pneumatophorus	68
pneumatophorus, Scomber	68
Podothecus	113
acipenserinus (1381)	114
decagonus (1379)	114
vulsus (1380)	114
Pœcilia couchiana (592)	50
Pœcilichthys	80
asprigenis	81
beani	78
borealis	80
butlerianus	81
camurus	80
eos	81
jessiæ	81
palustris	81
punctulatus	80, 81
quiescens	81
sagitta	80
sanguifluus	80
swaini	81
vulneratus	80
zonalis	80
pœcilopus, Myripristis (837)	75, 76
Rhamphoberyx	76
pœcilurum, Moxostoma (191)	20
poëyi, Hemirhamphus	60
Pogonias chromis (1084)	93
Pogonichthys argyriosus	30
macrolepidotus (350)	30
symmetricus	32
polaris, Cottus	110
Lycodalepis (1518)	125
Poliotrema dombeyi (3)	3
politus, Scriphus (1121)	95
Tetrodon (1670)	140
Pollachius chalcogrammus (1562)	130
saida (1563)	130
virens (1561)	130
pollicaris, Uranidea (1324)	111
polyacanthocephalus, Cottus (1337)	111
polyactocephalus, Blennius	122
Chirolophus (1470)	122
Polydactylus plumieri	66
polylepis, Balistes (1661)	140
Sebastes	108
Polynemidæ (Family LXXVIII)	66
Polynemus approximans (744)	66
melanopoma	66
octofilis	66
octonemus (746)	66
opercularis (745)	66
virginicus (743)	66
Polyodon spathula (100)	13
Polyodontidæ (Family XXVI)	13
Polyprion americanus (974)	83
Pomacanthodes	103

	Page.
Pomacanthus arcuatus	103
aureus (1207)	103
balteatus	103
crescentalis	103
zonipectus (1206)	103
Pomacentridæ (Family CXIV)	101
Pomacentrus analigutta	102
caudalis (1186)	101
flavilatus (1188)	102
leucostictus (1185)	101
obscuratus (1184)	101
quadrigutta (1189)	102
rectifrænum (1187)	102
rubicundus (1190)	102
Pomadasys axillaris (1030)	88
branicki (1032)	89
cæsius	89
cantharinus	88
elongatus (1028)	88
inornatus	88
leuciscus (1027)	88
macracanthus (1033)	89
nitidus (1029)	88
panamensis (1031)	89
Pomatomidæ (Family LXXXVI)	72
Pomatomus saltatrix (814)	72
Pomatopriou bairdii	102
Pomolobus	36
Pomotis aquilensis	77
pallidus	77
pomotis, Acantharchus (847)	76
Pomoxys annularis (842)	76
sparoides (843)	76
ponderosus, Amiurus (133)	15
Porichthys margaritatus (1420)	116
notatus	116
plectrodon	116
porosissimus (1421)	116
Poromitra capito (832)	75
Poronotus	73
porosissimus, Porichthys (1421)	116
Potamocottus	111
Potamorrhaphis	50
powelli, Balistes (1660)	140
præcisus, Eumesogrammus (1484)	122
pretiosus, Hypomesus (500)	42
Priacanthidæ (Family CIII)	86
Priacanthus arenatus	86
catalufa (1000)	86
macrophthalmus	86
prieto, Lutjanus	87
princeps, Caulolatilus (1215)	104
Prionistius macellus (1355)	112
Prionodes	82
fasciatus	83
Prionotus alatus (1386)	114
evolans (1390)	115
lineatus	115
palmipes (1385)	114
punctatus	114
ophryas (1387)	115
sarritor	115
scitulus (1384)	114
stearnsi (1388)	115
stephanophrys (1392)	115

CATALOGUE OF THE FISHES OF NORTH AMERICA.

	Page.
Prionotus strigatus (1391)	115
tribulus (1389)	115
Prionurus microlepidotus	103
punctatus (1211)	103
Pristididæ (Family XVIII)	10
Pristigaster lutipinnis	37
tartoor	37
Pristipoma brevipinne	88
fasciatum	88
fulvomaculatum	88
notatum	89
Pristis pectinatus (56)	10
perrottetii (57)	10
Proarthri	4, 5
probatocephalus, Diplodus (1066)	91
proboscideus, Chænomugil (719)	64
procerum, Nettastoma (634)	54, 55
procne, Notropis (234)	23
productus, Alepocephalus (428)	34
Eucinostomus	96
Merlucius (1567)	131
Rhinobatus (58)	10
prœliaris, Alvarius (944)	81
profundorum, Leptophidium (1530)	126
prolixum var. (196 b)	19
promelas, Pimephales (218)	22
Promicrops guasa	84
itaiara (976)	84
Promicropterus	86
decoratus	86
Pronotogrammus multifasciatus	83
proridens, Calamus (1055)	90
proriger, Sebastichthys (1270)	107
Sebastodes	107
proserpina, Moniana	25
Notropis (250)	25
Prosopium	43
Prospinus	84
prosthemius, Ceratichthys	29
prosthistius, Amiurus	15
Protoporus	27
proximus, Microgadus (1559)	130
Psettichthys melanostictus (1609)	135
Pseudarius	15
pseudogula, Eucinostomus	95
pseudohispanica, Clupea (441)	30
Pseudojulis modestus (1165)	99
notospilus (1164)	99
Pseudopleuronectes	136
Pseudopriacanthus altus (1001)	86
Pseudopristipoma	89
Pseudorhombus dentatus	134
oblongus	134
ocellaris	134
Pseudoscarus	100
Pseudotriacis microdon (22)	6
Pseudotriakis microdon	6
Psilonotidæ	141
Psilonotus punctatissimus (1676)	141
psittacus, Coryphæna	100
Scarus	100
Xyrichthys (1168)	100
Psychrolutes paradoxus (1302)	109
Pteraclis carolinus (823)	73
Pterophrynoides histrio (1640)	138

	Page.
Pteroplatea crebripunctata (82)	11
maclura (83)	11
marmorata (84)	11
Ptilichthyidæ (Family LXVI)	58
Ptilichthys goodei (650)	58
Ptychochilus harfordi (356)	30
lucius (357)	30
oregonensis (364)	30
rapax (355)	30
Ptychostomus albidus	19
puellaris, Cossyphus	98
Decodon (1156)	98
pugetensis, Chitonotus (1311)	110
pulchella, Liparis (1398)	115
Moniana	24
pulchellus, Phoxinus (388)	31
pulcher, Phoxinus (380)	31
Trochocopis (1157)	98
pullus, Monacanthus (1666)	140
pulverulentum, Oxygeneum (198)	20
punctata, Coryphæna	73
Opisthognathus (1436)	118
punctatissimus, Psilonotus (1676)	141
Tetrodon	141
punctatus var.	85
Apsonodon	8
Decapterus (777)	69
Dermatolepis (995)	85
Hypsoblennius (1445)	119
Ictalurus (134)	15
Iscsthes	119
Lepomis (861)	77
Micristodus (54)	10
Myrophis (630)	54
Prionotus	114
Prionurus (1211)	103
Squalus	8
Stichæus (1486)	122
punctifer, Ophichthys	53
punctifera, Dionda (207)	21
punctipinne, Siphostoma (682)	61
punctipinnis, Chromis (1193)	102
punctulata, Coryphæna	73
Uranidea (1318)	111
punctulatum, Etheostoma (933)	80
punctulatus, Alvarius (945)	81
Hippocampus (697)	62
Notropis (298)	27
Pœcilichthys	80, 81
pungitius, Pygosteus (707)	63
purpuratus, Salmo (525)	44
purpureum, Thalassoma	99
purpureus, Julis	99
Phoxinus (387)	31
Pusa radiata	99
putnami, Cottogaster (896)	78
pygmæa var. (506 b)	50
Pygosteus pungitius (707)	63
brachypoda (707 c)	63
concinnus (707 b)	63
pyrrhomelas, Notropis (371)	25
quadracus, Apeltes (714)	63
quadricornis, Cottus (1340)	111
Ostracion	139
quadrifasciatus, Chasmodes (1440)	119

	Page.		Page.
quadrifilis, Cottus (1346)	111	reticulatus, Monochir	137
quadrigutta, Pomacentrus (1180)	102	Otolithus	95
quadrilateralis, Coregonus (505)	43	retifer, Scylliorhinus (22)	5
quadriloba, Rhinoptera (96)	12	retifera, Muræna (604)	51
quadripinnis, Salarias	120	retropinnis, Catostomus (150)	17
quadriseriatus, Icelinus (1300)	110	Microdesmus	126
quadrituberculatus, Pleuronectes (1622)	136	retrosella, Amia	92
quadrocellata, Ancylopsetta (1601)	134	Apogon (1075)	92
Quassilabia lacera (194)	20	Rhacochilus toxotes (1148)	87
Querimana gyrans (721)	64	Rhamphoberyx leucopus	70
harengus (720)	64	pœcilopus	76
quiescens, Etheostoma (940)	81	Rhamphocottus richardsoni (1368)	113
Pœcilichthys	81	Rhegnopteri	66
radialis, Centropristis	82	rhessodon, Gobiesox (1414)	116
Serranus (962)	82	rhina, Raia (73)	11
radians, Sparisoma (1174)	100	Rhinichthys atronasus (321)	28
radiata var. (1651 b)	139	cataractæ (320)	27, 28
Malthe	8	dulcis (320 b)	27
Pnsa	99	transmontanus (320 c)	28
Raia (65)	11	dulcis	28
radiatus, Labrus	98	transmontanus	28
Lophius	139	Rhinobatidæ (Family xix)	10
Platyglossus (1158)	98	Rhinobatus exasperatus (61)	10
Sparus	98	glaucostigma (59)	10
Raia ackleyi ornata (67)	11	lentiginosus (60)	10
binoculata (74)	11	productus (58)	10
eglanteria (66)	11	triseriatus (62)	10
erinacea (63)	11	Rhinodontidæ (Family xvi)	10
granulata (69)	11	Rhinogobius	105
inornata (72)	11	Rhinonemus cimbrius (1537)	128
inermis (72 b)	11	Rhinoptera quadriloba (96)	12
lævis (75)	11	Rhinotriacis	7
ocellata (64)	11	rhodochloris, Sebastichthys (1280)	108
parmifera (70)	11	rhodopus, Trachynotus (798)	71
plutonia (68)	11	rhodorus, Ascelichthys (1301)	109
radiata (65)	11	rhodoterus, Holconotus (1141)	96
rhina (73)	11	rhomaleus, Gnathypops (1431)	118
stellulata (71)	11	Opisthognathus	118
Raia	10	Squalus	31
Raiidæ (Family xx)	11	rhombeus, Gerres	96
raji, Brama (824)	73	Rhombochirus osteochir (755)	66
ranula, Liparis (1402)	115	rhomboides, Diplodus (1064)	91
Ranzania truncata	141	Trachynotus (800)	71
rapax, Ptychochilus (355)	30	Rhomboidichthys leopardinus	132
rastrelliger, Sebastichthys (1285)	108	Rhomboplites	87
rectifrænum, Pomacentrus (1187)	102	aurorubens (1019)	88
regale, Cynoscion (1113)	95	Rhombus	72
regalis, Scomberomorus (768)	68	ocellatus	132
regia var. (216 c)	22	rhothea, Uranidea (1316)	110
regius, Hybognathus	21	Rhothœca	80
Phycis (1543)	129	Rhypticidæ (Family cii)	85
reinhardti, Careproctus (1396)	115	Rhypticus bistrispinus (998)	86
Himantolophus (1650)	139	maculatus	86
Onos (1538)	128	nigripinnis (999)	86
Reinhardtius hippoglossoides (1592)	133	pituitosus	86
Remora albescens (754)	66	saponaceus (996)	85
brachyptera (753)	66	xanti (997)	85
remora (752)	66	ricei, Uranidea (1313)	110
remora, Remora (752)	66	richardi, Hemirhamphus	60
Reniceps	8	richardsoni, Rhamphocottus (1368)	113
reticulata, Solen	137	Uranidea (1320)	111
reticulatum, Cynoscion (1119)	95	Richardsonius balteatus (419)	33
reticulatus, Chilomycterus (1682)	141	lateralis (420)	33
Esox (599)	50	rimator, Hæmulon (1043)	89
Lycodes (1512)	124	ringens, Stolephorus (457)	37

[177] CATALOGUE OF THE FISHES OF NORTH AMERICA.

	Page.
ringens, Sudis (475)	38
riverendi, Cyprinodon (540)	47
Trifarcius	47
rivoliana, Seriola (809)	72
rivulatus, Cirrhites (1072)	92
Cirrhitichthys	92
robalito, Centropomus (953)	82
roberti, Exocœtus	61
Hemirhamphus (666)	60
robusta, Gila (359)	30
Roccus americanus (957)	82
chrysops (955)	82
interruptus (956)	82
lineatus	82
septentrionalis (954)	82
Roncador stearnsi (1085)	93
roncador, Umbrina (1101)	94
rondeleti, Exocœtus (674)	61
rosacea, Mycteroperca (977)	84
rosaceus, Brachyistius (1136)	96
Sebastichthys (1279)	108
rosæ, Hemirhamphus (667)	60
roseipinnis, Notropis (299)	27
roseus, Cryptotomus	100
Notropis (277)	26
rosipes, Xyrichthys (1171)	100
rostrata var. (638)	55
Anguilla	55
Antimora	129
rostratus, Brachyopsis (1374)	113
Haloporphyrus	129
Heterostichus (1463)	121
Tetrodon	141
rothrocki, Notogrammus (1487)	123
rubellus, Alburnus	27
ruber var. (994)	85
Sebastichthys (1276)	108
rubicundus, Acipenser (104)	13
Pomacentrus (1190)	102
rubricroceus, Notropis (278)	26
rubrifrons var. (331 b)	29
Notropis (310)	27
Zygonectes (571)	49
rubripinna, Cyprinella	25
rubripinnis, Argyreus	27
Minnilus	27
rubrovinctus, Sebastichthys (1283)	108
rufolineatum, Etheostoma (922)	80
rufus, Bodianus (1153)	97
Labrus	97
Opos (1540)	128
rupestre, Etheostoma (929)	80
rupestris, Ambloplites (845)	76
Coryphænoides (1573)	131
Macrurus	131
Xiphister (1482)	122
Rupiscartes atlanticus (1455)	120
chiostictus (1454)	120
rutila, Moniana	24
Rutilus storerianus	24, 28
sabina, Trygon (91)	12
saburræ, Chasmodes (1441)	119
Saccopharyngidæ (Family LXIV)	57
Saccopharynx ampullaceus (648)	57
flagellum	57

	Page.
sacer, Anthias	83
sagax, Clupea (440)	36
sagitta, Etheostoma (927)	80
Pœcilichthys	80
Tylosurus (659)	59
Tyntlastes (1249)	106
sagittula, Enctenogobius	105
Gobius (1220)	105
saida, Pollachius (1563)	130
Salar	44
salar, Salmo (523)	44
Salarias alticus	120
atlanticus	120
chiostictus	120
quadripinnis	120
Salmo gairdneri (524)	44
irideus (524 b)	44
irideus	44
mülleri	40
purpuratus (525)	44
bouvieri (525 b)	44
henshawi (525 d)	44
spilurus (525 e)	44
stomias (525 c)	44
salar (523)	44
sebago (523 b)	44
salmoides, Micropterus (876)	77
salmoneus, Chanos	35
Esox	50
Mugil	35
Salmonidæ (Family XLIX)	42, 43
saltatrix, Pomatomus (814)	72
saludanus, Alburnops	24
Salvelinus arcturus (528)	44
fontinalis (530)	44
immaculatus (530b)	44
malma (529)	44
namaycush (526)	44
siscowet (526 b)	44
naresi	44
nitidus	44
oquassa (527)	44
stagnalis (531)	44
sanguifluus, Pœcilichthys	80
sanguineus, Antennarius (1643)	138
sapidissima, Clupea (446)	36
saponaceus, Anthias	85
Rhypticus (906)	85
sara, Cybium	68
Sarda chilensis (772)	69
sarda (771)	69
sarda, Sarda (771)	69
sardina, Clupea (447)	36
Harengula	36
Sardinia	36
Sargus caribæus	91
unimaculatus	91
Sarothrodus nigrirostris	102
sarritor, Prionotus	115
satiricus, Neoclinus (1457)	120
saturnus, Johnius (1092)	93
Saurida	39
Saurus anolis	39
intermedius	39
spixianus	39

	Page.
saurus, Elops (433)	34
Oligoplites (813)	72
Scomberesox (663)	60
saxatile, Etheostoma (928)	80
saxatilis, Glyphidodon (1192)	102
Johnius	94
Menticirrus (1108)	94
Perca	82
sayanus, Aphredoderus (838)	70
sayi, Trygon (87)	12
scaber, Hexagrammus (1254)	107
Uranoscopus	117
scabriceps, Notropis (287)	26*
Scaphirhynchops platyrhynchus (106)	13
scaphiura, Opisthognathus (1434)	118
Scarus	100
cœruleus (1179)	101
croicensis (1178)	101
guacamaia (1180)	101
perrico (1181)	101
psittacus	100
squalidus	101
scepticus, Notropis (304)	27
Schedophilopsis spinosus	104
Schedophilus	104
Schilbeodes	14
schlegeli, Lotella	130
schneideri, Ophichthys (626)	53
schœpfi, Alutera (1607)	140
Sciadarius	15
sciadicus, Zygonectes (575)	49
Sciæna acuminata	94
chrysura (1087)	93
icistia (1088)	93
jacobi (1089)	93
lanceolata (1080)	93
lineata	82
ocellata (1091)	93
sciera (1090)	93
Sciænidæ (Family CIX)	93
sciera, Sciæna (1090)	93
scierus, Hadropterus (913)	79
scituliceps, Synodus (479)	39
scitulus, Prionotus (1324)	114
sciurus, Hæmulon (1047)	90
Sparus	90
Sclerognathus	16
Scoliodon	7, 8
terræ-novæ	8
scolopaceus, Nemichthys (642)	56
scolopax, Macrorhamphosus (701)	62
Scomber colias (763)	68
pneumatophorus	68
scombrus (764)	68
speciosus	70
Scomberesocidæ (Family LXVIII a)	59, 60
Scomberesox brevirostris (664)	60
saurus (663)	60
Scomberomorus cavalla (769)	68
concolor (766)	68
maculatus (767)	68
regalis (768)	68
Scombridæ (Family LXXXIV)	68
Scombroidinæ Family (LXVIII a)	59, 60
scombrus, Scomber (764)	68

	Page.
Scopelidæ (Family XLV)	39, 40, 42
Scopelus boops	40
mülleri	40
Scorpæna	108
brasiliensis (1297)	109
bufo	8
calcarata	109
grandicornis (1296)	109
guttata (1294)	109
occipitalis (1298)	109
plumieri (1295)	8, 109
stearnsi	109
Scorpænichthys marmoratus (1361)	112
Scorpænidæ (Family CXXIII)	107
scorpioides, Cottus (1335)	111
scorpis californiensis	92
georgianus	92
scorpius, Cottus (1336)	111
scripta, Alutera (1608)	110
scrutator, Hypsoblennius (1447)	119
scudderi, Hæmulon (1050)	90
scuticaris, Callechelys (614)	52
scylla, Notropis (236)	24
Scylliidæ (Family IX)	5
Scylliorhinus retifer (22)	5
ventriosus (21)	5
Scymnidæ (Family VII)	5
Scytalina	126
Scytaliscus cerdale (1523)	120
sebago var. (523 b)	44
Sebastes kuhli	108
marinus (1262)	107
polylepis	108
Sebastichthys atrovirens (1272)	107
auriculatus (1284)	108
brevispinis (1271)	107
carnatus (1288)	108
chrysomelas (1286 b)	108
caurinus (1286)	108
vexillaris (1286 b)	108
chlorostictus (1281)	108
ciliatus (1266)	107
constellatus (1278)	108
elongatus (1282)	108
entomelas (1268)	107
flavidus (1264)	107
maliger (1287)	108
matzubaræ (1275)	107, 108
melanops (1265)	107
miniatus (1274)	108
mystinus (1267)	107
nebulosus (1289)	108
nigrocinctus (1291)	108
ovalis (1269)	107
pinniger (1273)	107, 108
proriger (1270)	107
brevispinis	107
rastrelliger (1285)	108
rhodochloris (1280)	108
rosaceus (1279)	108
ruber (1276)	108
rubrovinctus (1283)	108
serriceps (1290)	108
umbrosus (1277)	108
variabilis	107

CATALOGUE OF THE FISHES OF NORTH AMERICA.

	Page.
Sebastodes paucispinis (1263)	107
proriger	107
Sebastomus	108
Sebastoplus dactylopterus (1293)	108
Sebastopsis xyris (1292)	108
Sebastosomus	107
sectatrix, Kyphosus (1069)	92
Perca	92
seemanni, Arius	15
Galeichthys (137)	15
selachops, Apterichthys	52
Ichthyapus (612)	52
Selachostomi	13
Selene	69
œrstedi (792)	71
vomer (793)	71
selene, Luxilus	24
selenops, Hyodon (432)	34
sellicauda, Epinephelus (985)	84
semicinctus, Platyglossus (1162)	90
semicoronata, Seriola	72
semifasciatus, Triacis (28)	7
seminolis, Fundulus (561)	45
seminuda, Gila (365)	30
seminudus, Lycodes (1513)	124
semiscabra, Uranidea (1315)	110
Semotilus atromaculatus (347)	29
bullaris (349)	29
diplæmius	26
thoreauianus (348)	29
senilis, Gambusia (500)	50
senticosa, Halientæa (1654)	139
septentrionalis, Motella	128
Onos (1541)	128
Perca	82
Roccus (954)	82
serena, Dionda (211)	21
Seriola	69
aliciola	72
dorsalis (807)	72
dumerili (805)	71, 72
lalandi (805 b)	71
falcata	72
fasciata (808)	72
lalandi	71, 72
mazatlana (806)	72
rivoliana (809)	72
semicoronata	72
zonata (804)	71
carolinensis (804 b)	71
Seriolinæ	69
Seriphus politus (1121)	95
serpentinus, Leptoblennius (1493)	123
Serranidæ (Family CI)	82, 85, 86
Serranus arara	84
atrarius (958)	82
bonaci	84
brunneus	84
caloptcryx (965)	83
clathratus (966)	83
formosus (961)	82
furvus (959)	82
itaiara	84
maculofasciatus (967)	83
nebulifer (968)	83

	Page.
Serranus petrosus	84
philadelphicus (960)	82
phœbe (964)	82
radialis (962)	82
subligarius (963)	82, 83
Serraria	79
serrata, Fistularia (703)	63
serriceps, Sebastichthys (1290)	108
serrifer, Conodon (1021)	88
Serrivomer beani (647)	57
sessilicauda, Monolene (1630)	136
Setarches parmatus (1299)	109
setipinnis, Vomer (791)	71
sexfasciatum, Hæmulon (1053)	90
shufeldti, Typhlopsaras	138
shumardi, Alburnops	23
Cottogaster (898)	79
Siboma	31
sicculus, Labidesthes (728)	65
Sidera castanea (606)	51
dovii (608)	51
funebris (610)	52
mordax (607)	51
moringa (611)	52
ocellata (609)	51
siderium, Zophendum (204)	20
sierrita, Tylosurus	50
Sigmops stigmaticus (538)	46, 47
signatus, Bathymaster (1213)	104
siguifer, Chatoëssus	36
Stypodon (351)	30
Thymallus (516)	43
Siluridæ (Family XXX)	14
Silurus catus	14
Simenchelyidæ (Family LXI)	56
Simenchelys	52
parasiticus (639)	56
similis, Fundulus (558)	45
simillimus, Stromateus (818)	73
simotera, Ulocentra (891)	78
simoterum, Diplesion	78
simula, Chaliuura (1575)	132
simulans, Enneacanthus (851)	76
simus, Notropis (245)	24
Siphagonus barbatus (1373)	113
Siphateles vittatus	32
Siphostoma affine (690)	61, 62
auliscus (685)	61
bairdianum (687)	61
barbaræ (686)	61
californiense (683)	61
crinigerum (694)	62
floridæ (689)	62
fuscum (692)	62
griseolineatum (684)	61
leptorhynchum (688)	62
louisianæ (691)	62
mackayi (693)	62
punctipinne (682)	61
zatropis (681)	61
siscowet var. (526 b)	44
sloani, Chauliodus (536)	46
smaragdus, Eleotris	104
Erotelis (1210)	104
socius, Notropis (293)	26

	Page.
solandri, Acanthocybium (770)	68
Cybium	68
Solea inscripta	137
mazatlana	137
pilosa	137
reticulata	137
Soleidæ (Family CXLVIII)	137
Somniosus microcephalus (17)	5
soporator, Gobius (1228)	105
sordidus, Citharichthys (1583)	133
Sparidæ (Family CV)	86, 92
Sparisoma cyanolene (1176)	101
flavescens (1177)	101
radians (1174)	100
xystrodon (1175)	101
sparoides, Pomoxys (843)	76
Sparus abildgaardii	101
argyrops	91
caxis	87
chrysops	91
pagrus (1054)	90
radiatus	98
sciurus	90
spathula, Polyodon (100)	13
spatula, Lepidosteus	13
speciosus, Caranx (788)	70
Scomber	70
spectabile var. (936 b)	81
spectrunculus, Notropis (228)	23
spelæus, Amblyopsis (539)	47
spengleri, Tetrodon (1672)	141
spet, Sphyræna	65
Sphagebranchus	52
Sphyræna argentea (738)	65
borealis (739)	65
ensis (742)	65
forsteri	65
guaguanche (740)	65
picuda (741)	65
spet	65
Sphyrænidæ (Family LXXVII)	65
Sphyrna tiburo (45)	8, 9
tudes (46)	9
zygæna (47)	9
Sphyrnidæ (Family XI)	8
spilonotus, Monacanthus (1665)	140
spilopterus, Citharichthys (1585)	133
spilota, Uranidea (1323)	111
spilurus var. (525 e)	44
Notropis (290)	26
Spinacidæ (Family VIII)	5
Spiniromer goodei (646)	57
spinosus, Echinorhinus (10)	5
Eumicrotremus (1409)	116
Hemilepidotus (1357)	112
Schedophilopsis	104
spixianus, Saurus	39
Synodus (478)	39
Spratelloides bryoporus	35
Squali	5
squalidus, Scarus	101
Squalius	27, 30
galtlæ	31
lemmoni	31
rhomaleus	31

	Page.
Squalus acanthias (19)	5
canis	6
carcharias	8
cœrulous	8
galeus	6
longimanus	8
mustelus	6
punctatus	8
squamatus, Hadropterus (909)	79
Phoxinus (396)	31
squamiceps, Etheostoma (925)	80
squamilentus, Cousius (342)	29
Paralichthys (1590)	134
Squatina squatina (55)	10
squatina, Squatina (55)	10
Squatinidæ (Family XVII)	10
stagnalis, Salvelinus (531)	44
stearnsi, Blennius (1450)	110
Lutjanus	87
Prionotus (1388)	115
Roncador (1085)	93
Scorpæna	109
steindachneri, Diabasis	90
Hæmulon (1048)	90
stellatus, Pleuronectes (1621)	136
stelleri, Cyclopterichthys (1408)	116
stellifer, Fundulus (570)	49
Stelliferus	93
stellulata, Raia (71)	11
Stenodus leucichthys	43
mackenziei (517)	43
Stenotomus aculeatus	91
caprinus (1062)	91
chrysops (1063)	91
aculeatus (1063b)	91
Stephanoberyx monæ (828)	74
stephanophrys, Prionotus (1392)	115
Stereolepis gigas (975)	83
sterletus, Ceratichthys	29
Sternoptyx diaphana (535)	45, 46
olfersi	45
Sternoptychidæ (Family LI)	40, 45
Stichæinæ	123
Stichæus punctatus (1486)	122
stigma, Gymnelis	125
stigmæa, Ulocentra (890)	78
stigmæus, Citharichthys (1584)	133
stigmaticus, Gobionellus (1236)	106
Signiops (538)	46, 47
stigmaturus var. (260 b)	25
Gobius (1231)	105
Photogenis	25
stilbius, Notropis (307)	27
stipes, Atherina (725)	65
Stizostedion canadense (949)	81
boreum (949 c)	81
grisoum (949 b)	81
vitreum (948)	81
Stoasodon laticeps (93)	12
narinari (92)	12
Stolephorus browni (460)	37, 38
compressus (471)	38
curtus (465)	38
delicatissimus (469)	38
eurystolo (464)	38

CATALOGUE OF THE FISHES OF NORTH AMERICA.

	Page.
Stolephorus exiguus (467)	38
ischanus (462)	38
lucidus (470)	38
macrolepidotus (458)	37
miarchus (468)	38
mitchilli (466)	38
opercularis (459)	37
perfasciatus (463)	37, 38
perthecatus (461)	37
ringens (457)	37
stolifera, Clupea (450)	36
Dussumieria (430)	35
stolzmanni, Belone	59
Tylosurus (662)	59
Stomias ferox (489)	41
stomias var. (525 c)	44
Atheresthes (1593)	133
Stomiatidæ (Family XLVII)	41
storerianus, Hybopsis (330)	28
Rutilus	28
stramineus var. (233 b)	23
Notropis	23
striatus, Blennius	119
Epinephelus (984)	84
strigatus, Antennarius (1644)	138
Holacanthus (1204)	103
Prionotus (1391)	115
Stromateidæ (Family LXXXVIII)	72
Stromateus medius (817)	73
paru (816)	72
simillimus (818)	73
triacanthus (819)	73
strumosus, Gobiesox (1412)	116
sturio, Acipenser (101)	13
stylifer, Hippocampus (699)	62
Stypodon signifer (351)	30
suavis, Cyprinella	24
subbifurcatus, Eumesogrammus (1485)	122
subligarius, Serranus (963)	82, 83
suborbitale, Holocentrum (835)	75
suborbitalis, Plectromus (830)	74
subterraneus, Typhlichthys (540)	47
sucetta, Catostomus	19
Erimyzon (170)	19
Sudis borealis (476)	38
coruscans	38
ringens (475)	38
sueuri, Coryphæna	73
Suillus	97
suillus, Lachnolæmus	97
superciliosus, Hexagrammus (1255)	107
Hyborhynchus	22
surinamensis, Lobotes (1002)	86
ansanæ, Boleosoma (887)	78
swaini, Notropis (294)	26
Pœcilichthys	81
swampina, Fundulus	48
swani, Bothragonus (1377)	114
symmetrica, Alganaea (407)	32
symmetricus, Lepomis (854)	77
Pogonichthys	32
Symmetrurus argyriosus	30
synagris, Lutjanus (1012)	87
Synaphobranchidæ (Family LXII)	56
Synaphobranchus pinnatus (640)	56

	Page.
Synentognathi	59
Syngnathidæ (Family LXIX)	61, 62
Syngnathus bairdianus	61
Synodontidæ (Family XLIV)	39
Synodus	40
anolis (481)	39
cubanus	39
fœtens (477)	39
intermedius	39
lucioceps (480)	39
myops (482)	39
scituliceps (479)	39
spixianus (478)	39
syrtensium, Argentina (502)	42
tabacaria, Fistularia (702)	63
tænia, Phoxinus (371)	30
tæniatum, Hæmulon (1044)	90
tæniatus var. (1039 b)	80
tæniatus, Anisotremus	89
tæniops, Eucacentrus (993)	85
tæniopterus, Cottus (1339)	111
Tæniotoca	96
taboensis, Catostomus (161)	17
tanneri, Hyperchoristus (490)	41, 42
tartoor, Pristigaster	37
tau, Batrachus (1419)	116
Taurida	110
taurocephalus, Alburnops	22
taurus, Carcharias	9
Odontaspis	7
Tautogolabrus	97
taylori, Otophidium (1528)	126
tchawytcha, Oncorhynchus (520)	44
telescopus, Notropis (306)	27
telfairi, Agonostomus	64
tenuifilis, Antennarius	138
tenuis, Leuresthes (727)	65
Phycis (1547)	129
teres, Callechelys (615)	52, 53
Catostomus (170)	18
Etrumeus (437)	35
teretulus, Phenacobius (315)	27
tergisus, Hyodon (431)	34
terræ-novæ, Carcharhinus (44)	8
Scoliodon	8
tessellatus, Hadropterus (914)	79
Tetrodon (1671)	140, 141
Tetragonopterus argentatus (425)	34
Tetrapturus albidus (758)	67
tetraspilus, Upeneus	93
Tetrodon annulatus	141
heraldi	141
lineatus	140
nephelus (1673)	141
oxyrhynchus	141
politus (1670)	140
punctatissimus	141
rostratus	141
spengleri (1672)	141
testudineus (1671)	140, 141
annulatus (1671 b)	141
trichocephalus (1675)	141
turgidus (1674)	141
Tetrodontidæ (Family CLV)	140
Teuthis cœruleus (1210)	103

	Page.
Teuthis hepatus (1208)	103
tractus (1209)	103
texana, Anguilla	55
texensis, Dionda	21
thalassina, Algansea (410)	32
thalassinum, Cynoscion (1114)	95
Etheostoma (918)	80
Moxostoma (183)	19
thalassinus, Doratonotus (1167)	99
Lepidogobius (1241)	106
Myloleucus	32
Thalassoma lucasanum (1166)	99*
purpureum	99
Thaleichthys pacificus (496)	42
thaleichthys, Osmerus (497)	42
thazard, Auxis (765)	68
thompsoni, Triglopsis (1350)	112
thoreauianus, Semotilus (348)	29
thrissa, Clupea	36
thrissina, Clupea (448)	36
thryza, Clupea	36
Thymallus gymnothorax	43
ontariensis	43
signifer (516)	43
ontariensis (516 b)	43
tricolor	43
thynnus, Orcynus (774)	69
Thyris pellucidus	136
Tiaroga cobitis (319)	27
tiburo, Sphyrna (45)	8, 9
Tigoma	30
nigrescens	31
tigrinus, Galeocerdo	7
Myrichthys (626)	54
Tilesia	130
timpanogensis, Notropis (313)	27
tomcod, Microgadus (1560)	130
topeka, Chola	24
Notropis (242)	24
Torpedinidæ. (Family XXI)	11
Torpedo, californica (77)	11
occidentalis (76)	11
torvus, Cottunculus (1304)	110
toxotes, Rhacochilus (1148)	97
Trachinocephalus	39
Trachurops	69
crumenophthalmus (781)	70
Trachurus	69
alleiolus	72
fasciatus	72
picturatus (779)	70
trachurus (780)	70
trachurus, Trachurus (780)	70
Trachynotinæ	69
Trachynotus	69, 97
argenteus (797)	71
carolinus (796)	71
fasciatus (802)	71
glaucus (801)	71
goreensis	71
kennedyi (799)	71
nasutus	71
rhodopus (798)	71
rhomboides (800)	71
Trachypteridæ (Family CXVIII)	104

	Page.
Trachypterus altivelis (1212)	104
tractus, Teuthis (1209)	103
transmontanus var. (320 c)	28
Acipenser (102)	13
Rhinichthys	28
traski, Hysterocarpus (1132)	90
triacanthus, Stromateus (819)	73
Triacis henlei (29)	7
semifasciatus (28)	7
tribulus, Prionotus (1389)	115
Trichiuridæ (Family LXXXIII)	67
Trichiurus caudatus	67
lepturus (700)	67
trichocephalus, Tetrodon (1675)	141
Trichodiodon pilosus (1677)	141
Trichodon japonicus (1423)	117
trichodon (1422)	117
trichodon, Mugil (718)	64
Trichodon (1422)	117
Trichodontidæ (Family CXXXI)	117
trichroistius, Notropis (267)	25
tricolor, Holacanthus	103
Thymallus	43
tricorne, Ostracion (1657)	139
tricuspis, Gymnacanthus (1347)	112
tridentatus, Ammocœtes (4)	3
tridigitatus, Dactyloscopus (1426)	117
Trifarcius rivorendi	47
trifurca, Perca	82
Trigla evolans	115
lineata	115
Triglidæ (Family CXXVI)	114
Triglops pingeli (1354)	112
Triglopsis thompsoni (1350)	112
trigonum, Ostracion (1656)	139
tripteronotus, Blennius	121
Tripterygion carminale (1461)	121
triquetrum, Ostracion (1655)	139
triserialis, Ophichthys (620)	53
triseriatus, Rhinobatus (62)	10
Trisotropis	84
trispinosus, Odontopyxis (1378)	114
tristœchus, Lepidosteus (109)	13
Trochocopus pulcher (1157)	98
Tropidichthys	141
Tropidinius	87
troscheli var. (1192 b)	102
Gyphidodon	102
truncata, Ranzania	141
Trycherodon megalops	33
Trygon centrura (85)	12
dipterura (89)	12
hastata (86)	12
longa (88)	12
sabina (91)	12
sayi (87)	12
tuberculata (90)	12
Trygonidæ (Family XXII)	11
tuberculata, Trygon (90)	12
tudes, Sphyrna (46)	9
Zygæna	9
tuditanus, Hybopsis	22
Hypargyrus	22
tullibee, Coregonus (515)	43
tumidus var. (148 c)	10

CATALOGUE OF THE FISHES OF NORTH AMERICA.

	Page.
tunicata, Liparis (1400)	115
turgidus, Tetrodon (1674)	141
turneri, Lycodalepis (1517)	125
tuscumbia, Etheostoma (939)	81
Tylosurus caribbæus (657)	59
crassus (656)	59
exilis (661)	59
fodiator (655)	59
gladius	59
hians (654)	59
marinus (660)	59
notatus (658)	59
sagitta (659)	59
sierrita	59
stolzmanni (662)	59
Tyntlastes sagitta (1249)	106
Typhlichthys subterraneus (540)	47
Typhlogobius californiensis (1248)	106
Typhlopsaras shufeldti	138
tyrannus, Anguilla	55
Brevoortia (453)	37
Ulocentra atripinnis	78
blennius (893)	78
histrio (892)	78
phlox (889)	78
simotera (891)	78
stigmæa (890)	78
Umbra	51
limi (590)	50
pygmæa (596 b)	50
umbratilis, Alburnellus	26
Notropis (297)	27
Umbridæ (Family LV)	50
Umbrina analis	94
broussoneti (1104)	94
dorsalis (1103)	94
elongata	94
nasus	94
panamensis	94
roncador (1101)	94
xanti (1102)	94
umbrosa, Cyprinella	25
Narcine (79)	11
umbrosus, Esox	50
Sebastichthys (1277)	108
uncinatus, Artediellus (1212)	110
Cottus	110
undecimalis, Centropomus (950)	81
undulatus, Menticirrus (1107)	94
Micropogon (1099)	94
unicornis, Citharichthys (1588)	133
unifasciatus, Hemirhamphus (663)	60
unimaculata, Perca	91
unimaculatus, Diplodus (1065)	91
Sargus	91
univittatus, Apodicanthys (1478)	122
Upeneus balteatus	93
dentatus (1082)	93
flavovittatus	93
grandisquamis (1081)	93
maculatus (1079)	93
martinicus (1080)	93
tetraspilus	93
Upsilonphorus	118
guttatus (1429)	117
Upsilonphorus y-græcum (1428)	117
Uranidea aspera (1314)	110
bendirei (1319)	111
boleoides (1329)	111
cognata (1321)	111
formosa (1331)	111
franklini (1330)	111
gobioides (1328)	111
gracilis (1327)	111
gulosa (1317)	111
hoyi (1332)	111
marginata (1225)	111
minuta (1322)	111
pollicaris (1324)	111
punctulata (1318)	111
rhothea (1316)	110
ricei (1313)	110
richardsoni (1320)	111
alvordi (1320 e)	111
bairdi (1320 b)	111
carolinæ (1320 h)	111
kumlieni (1320 c)	111
meridionalis (1320 f)	111
wheeleri (1320 i)	111
wilsoni (1320 d)	111
zophera (1320 g)	111
semiscabra (1315)	110
spilota (1323)	111
viscosa (1326)	111
uranidea, Cottogaster (897)	79
uranops, Phenacobius (318)	27
Uranoscopidæ (Family CXXXIII)	117
Uranoscopus anoplos	118
scaber	117
y-græcum	117
uranoscopus, Mancalias (1647)	138
Uraspis	70
Urolophus asterias (81)	11
halleri (80)	11
Uronectes	125
urostigma, Cliola	25
urus, Ictiobus (145)	16
ustus, Cryptotomus (1172)	100
utowana, Catostomus	18
vafer, Myrophis (631)	54
vagrans, Menidia (730)	65
vahli, Lycodes (1510)	124
valenciennesi, Erotelis	105
Moxostoma (184)	19
vandoisulus, Phoxinus (367)	30
variabilis, Perca	108
Sebastichthys	107
variatum, Etheostoma	79
variatus, Alvordius	79
Hadropterus (912)	79
variegatus, Cyprinodon (545)	47
velatum, Moxostoma (179)	19
velicana, Atherina	65
velifer var	16
Ictiobus (148)	16, 17
Letharchus (613)	52
velox, Cliola	22
venenosa, Mycteroperca (981)	84
Perca	84
ventralis, Brosmophycis	127

REPORT OF COMMISSIONER OF FISH AND FISHERIES [181]

	Page.
ventralis, Dinematichthys (1533)	127
ventricosa, Apocope	28
ventricosus, Cyclopterichthys (1407)	116
ventriosus, Scylliorhinus (21)	5
venusta, Lucania (582)	49
venustus, Notropis (259)	25
Xyrichthys	100
veranyi, Cybium	68
Verilus	87
vermiculatus, Esox (598)	50
Xyrichthys	100
vernalis, Clupea (444)	36
verrilli, Lycenchelys (1509)	124
verrucosus, Brachyopsis (1375)	114
Cottus (1344)	111
verticalis, Pleuronichthys (1611)	135
vespertilio, Lophius	138
Malthe (1651)	139
vetula, Balistes (1658)	140
vetulus, Parophrys (1614)	135
vexillare, Boleosoma (886)	78
vexillaris var. (1286 b)	108
vigilax, Cliola (223)	22
vigilis, Ioa (884)	78
villosus, Malotus (495)	42
vinciguerræ, Exocœtus (675)	61
vinctus, Caranx (783)	70
Fundulus (568)	49
viola, Antimora (1530)	129
Haloporphyrus	129
violaceus, Cebedichthys (1483)	122
virens, Pollachius (1561)	130
virescens, Pantosteus	17
virgatulus, Gobiesox (1413)	116
virgatum, Etheostoma (926)	80
virgatus, Delolepis (1496)	123
virginicus, Anisotremus (1039)	89
Polynemus (743)	66
viridis, Gymnelis (1519)	125
viscosa, Uranidea (1326)	111
vitrea, Ioa (883)	78
vitreum, Stizostedion (948)	81
vittata, Alganæa (414)	32
Hemitremia	22
Lepidomeda (421)	33
vittatus, Siphateles	32
vivanus, Anthias (972)	83
Lutjanus (1013)	87
Mesoprion	87
vivax, Ammocrypta (881)	77
Cliola	22
volador, Exocœtus	61
volitans, Cephalacanthus (1393)	115
Exocœtus (676)	61
volucellus var. (233 d)	23
Hybopsis	23
Vomer	69
setipinnis (791)	71
vomer, Selene (793)	71
vulgaris, Amiurus (126)	15
vulnerata, Apocope	28
vulneratus, Pœcilichthys	80
vulpeculus	6
vulpes, Albula (429)	34
Alopias (48)	9

	Page.
vulsus, Podothecus (1380)	114
warreni, Boleichthys	81
webbi, Ophioblennius (1438)	119
wheatlandi var. (711 b)	63
wheeleri var. (1320 i)	111
whipplei, Etheostoma (934)	81
Notropis (261)	25
williamsoni, Coregonus (504)	43
Gasterosteus (709)	63
wilsoni var. (1320 d)	111
würdemanni, Gobius (1232)	105
xænocephalus, Notropis (284)	26
xænurus, Notropis (270)	25
xanthocephalus, Amiurus	14
xanthosticta var. (980 b)	84
xanthulum, Cynoscion (1118)	95
xanthurus, Liostomus (1095)	94
xanti var. (1459 b)	120
Clinus	120
Labrosomus	120
Rhypticus (997)	85
Umbrina (1102)	94
Xenichthys (1003)	86
Xenichthys xanti (1003)	86
xenops	86
xenicus, Fundulus	48
Xenisma	49
Xenistius californiensis (1004)	86
Xenomi	51
xenops, Xenichthys	86
Xiphias	68
gladius (757)	67
Xiphidium cruoreum	122
Xiphiidæ (Family LXXXII)	67, 68
Xiphister chirus (1480)	122
mucosus (1481)	122
rupestris (1482)	122
Xiphisterinæ	123
xyosternus, Brachyopsis (1376)	114
Xyrichthys lineatus	100
mundiceps (1169)	100
mundicorpus (1170)	100
pavo	100
psittacus (1168)	100
rosipes (1171)	100
venustus	100
vermiculatus	100
xyris, Sebastopsis (1292)	108
Xystreurys	134
liolepis (1603)	135
xystrodon, Sparisoma (1175)	101
Xystroplites	77
xysturus, Ophichthys	53
Ophisurus (618)	53
yarrelli, Phycis	129
y-græcum, Upsilonphorus (1428)	117
Uranoscopus	117
zachirus, Glyptocephalus (1627)	136
zanemus, Hybopsis (335)	29
Zaniolepis latipinnis (1258)	107
Zapteryx	10
zatropis, Siphostoma (681)	61
zebra var. (899 b)	79
zebra, Gobiesox (1416)	116
Pileoma	79

CATALOGUE OF THE FISHES OF NORTH AMERICA.

	Page.
zebrinus, Fundulus (560)	48
Zenidæ (Family XCIII)	74
Zenopsis ocellatus (827)	74
Zoarces anguillaris (1503)	124
zonale, Etheostoma (916)	80
zonalis, Pœcilichthys	80
zonata, Cliola	24
Seriola (804)	71
zonatum, Elassoma (839)	76
zonatus, Alburnus	26
Chætodipterus (1198)	102
Ephippus	102
Esox	49
Notropis (275)	26
zonifer, Clinus	120
Labrosomus (1460)	120
Myriolepis (1260)	107
Zygonectes (579)	49
zonipectus, Pomacanthus (1206)	103
zonistius, Notropis (276)	26
Zophendum plumbeum (205)	20
siderium (204)	20

	Page.
zophera var. (1320 g)	111
zophochir, Ophiohthys (625)	63
zosteræ Hippocampus (700)	*62
zosterurum, Gobiosoma (1245)	106
Zygæna tudes	9
zygæna, Sphyrna (47)	9
Zygonectes atrilatus	50
brachypterus	50
chrysotus (580)	49
cingulatus	49
craticula (578)	49
dispar (577)	49
floripinnis (573)	49
henshalli (572)	49
inurus	50
lineatus (574)	49
luciæ (581)	49
notatus (576)	49
rubrifrons (571)	49
sciadicus (575)	49
zonifer (579)	49
zyopterus, Galeorhinus (30)	7

ERRATA.

Species No. 8 should stand as *Petromyzon concolor*, Kirtland, instead of *P. bdellium*. *Ammocœtes concolor* seems to be the larva of this species.

Species 11 b. The subspecies should stand as *Petromyzon marinus unicolor* DeKay, instead of *P. m. dorsatus*. *Ammocœtes unicolor* DeKay is the larva of this form.

Genus 39. The name *Dasybatis* (Klein) Rafinesque, is prior to *Trygon* Adanson (1817), and must be used for this genus (cf. Garman, Proc., U. S. Nat. Mus., 1885).

Genus 61. *Hypentelium* should be reunited to *Catostomus*.

Species 328. Should stand as *Hybopsis kentuckiensis* Rafinesque, instead of *H. biguttatus*. It seems to be the *Luxilus kentuckiensis* Raf.

Species 601. Should apparently stand as *Esox masquinongy* Mitchill instead of *E. nobilior*.

The name of Family LXVIII a.—*Scomberesocidæ* was inadvertently omitted before genus 195, *Scomberesox*.

Species 1637 should apparently stand as *Aphoristia fasciata* Holbrook, instead of *A. plagiusa*.

www.ingramcontent.com/pod-product-compliance
Lightning Source LLC
Chambersburg PA
CBHW032147160426
43197CB00008B/797